MILLING MACHINE OPERATIONS

MILLING MACHINE OPERATIONS

Richard R. Kibbe

RRK General Machining and Manufacturing, Oxnard, California
Technology Division, Oxnard Community College, Oxnard, California

John Wiley & Sons

New York Chichester Brisbane Toronto Singapore

Library of Congress Catologing in Publication Data:

Kibbe, Richard R.
Milling machine operations.

Includes index.
1. Milling-machines. I. Title.
TJ1225.K49 1985 671.3′5 84-11798
ISBN 0-471-89020-0

Printed in the United States of America

10 9 8 7 6 5 4 3 2 1

Preface

This manual, which has been adapted from materials originally prepared by the Engineering Industry Training Board of Great Britain, presents many specific setups and operating procedures for milling machines. The step-by-step text is complemented by numerous illustrations. Review questions follow each major topic, and there is a glossary in the appendix. Because as much space as possible has been dedicated to specifics, this manual does not cover all the related background information that would be contained in a general text. Beginning or very inexperienced students should use this manual in conjunction with a general text on machine tools. *Milling Machine Operations* is most appropriate for educational institutions and industry training programs seeking expanded coverage of the subject in any easy-to-understand format. *Lathe Operations* and *Grinding Machine Operations* are also available in this series.

Richard R. Kibbe

Acknowledgments

The author thanks the following for their contributions toward the publication of this manual:

The Engineering Industry Training Board (EITB) of Great Britain

The individuals who reviewed the original materials and manuscript and made many useful comments and recommendations

John Wiley & Sons, Inc., for supplying additional illustrative materials

Contents

MILLING MACHINE OPERATIONS

General machine shop safety

Safety must be observed not only while you are in the machine shop, it must be observed in everything that you do. You must develop a safety attitude that extends to all your activities.

The machine shop is generally a safe and comfortable place to learn and work. However, the shop is full of potential hazards that you should be aware of so that you can protect yourself and your fellow workers. Developing a safety attitude and reviewing a few common sense safety rules will go a long way to prevent accidents in the shop.

Certainly no one wants an injury; but all of us sometimes take chances. If you knew ahead of time that you would be injured by engaging in a careless act, you would certainly avoid it. Developing a safety attitude means that you will not expose yourself to unreasonable risk because you understand the potential dangers. Remember that an injury can severely limit or end your career and leave you with a permanent physical handicap. This is hardly worth the risk of a brief instant of carelessness in the shop or while operating a machine.

Be aware of the following shop-related hazards and always keep yourself in a position to avoid accidents.

1 Wear your eye protection at all times in the shop and while operating the machine. Even when walking through the shop, beware of flying metal. Wear side shield glasses around grinding machines.

2 Dress properly for shop work, just as you would for any activity. This means appropriate shoes and shop coat or apron; also, keep loose clothing and hair out of danger. If your hair or clothing becomes tangled in a powerful moving mechanism on a machine tool, you will be pulled in and seriously injured. Such accidents can easily be avoided by being aware of potential dangers and protecting yourself.

3 Remove your rings and wristwatch. Like clothing, these items can be caught and pulled into the machine.

4 Most machine shops have compressed air delivered to each machine station, and you will see the air used to blow chips from the machine during cleanup. Although this practice is very common, it is also very dangerous. Some people do not realize just how much energy can be stored in compressed gas. The air blast propels the metal chips with great velocity. The chips can hit people across the shop, or they may be blown back into the face of the person at the machine. It is best to remove chips with a brush. If you are using air for this purpose, be extremely careful in the direction of the blast and blow at a greatly reduced pressure if possible. Do not blow off your clothing or skin with air. You are almost sure to blow chips and dirt in your eyes, or have dirt injected into your skin. Turn off the air supply at the valve before removing any hoses or disassembling any part of the air system.

5 Never handle chips with your bare hands. Chips usually are razor sharp and often are hot. Use a brush or a chip rake to pull chips from under the machine.

6 Keep the shop orderly by keeping the floor and machine clear of chips and scrap materials. Clean all spilled oil and coolant from the floor immediately. Spills make the floor slippery and extremely dangerous. Oil-soaked rags must be kept in an approved fireproof safety can that is emptied each day.

7 Carry long materials in the vertical position, or get help to carry a long or heavy piece of material. Be careful of light fixtures and ceilings.

8 Don't carry sharp tools in your pockets. You or someone else could be injured.

9 If you have an electrical problem with a machine, consult your instructor or supervisor. Do not open any electrical panels or switches. A high voltage shock can be fatal. If you are doing electrical service on a machine or in the shop, be sure that the service disconnect switch or circuit breaker is OFF and tagged with an appropriate warning.

10 Get help in lifting heavy tooling such as chucks and vises. Dropping a heavy object on a finger or hand can result in a permanent injury. If you are lifting alone, conform to proper lifting techniques.

11 Grinding dust, fumes, and smoke from welding or cutting must not be inhaled. These substances must be properly contained or vented from the shop.

12 Do not talk to others when they are operating a machine. If they are distracted there may be an accident.

13 Scuffling, horseplay, and practical joking are not tolerated in the shop. These can lead to an injury and you being fired from your job.

14 Be observant when you are in the shop. If you see a hazardous situation or a fellow student engaged in a careless or dangerous activity, a word of caution from you will surely be appreciated.

Since this manual covers a specific machine type, safety relative to that machine and its processes is covered in Unit 1.

UNIT ONE

Introduction to Milling Machine Operations

The machine tool

The milling machine is one of the most common and versatile machine tools. Common types include the following.

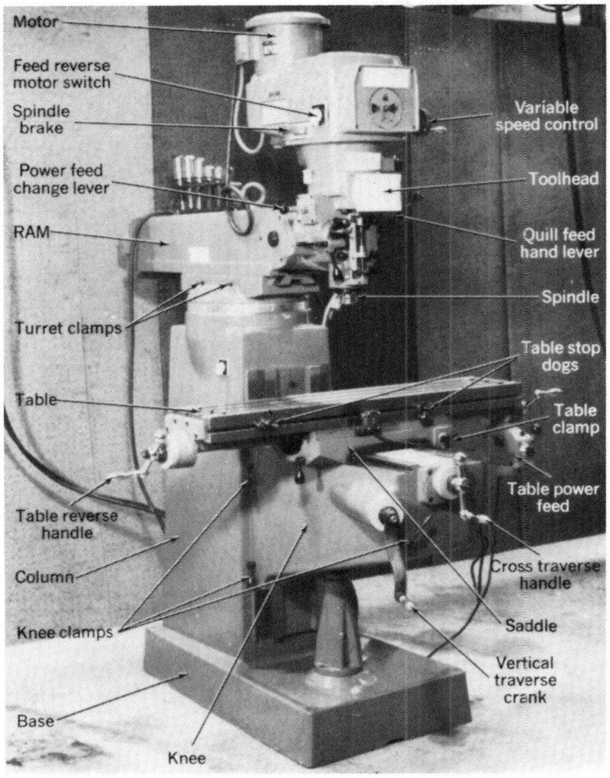

Knee and column vertical spindle mill

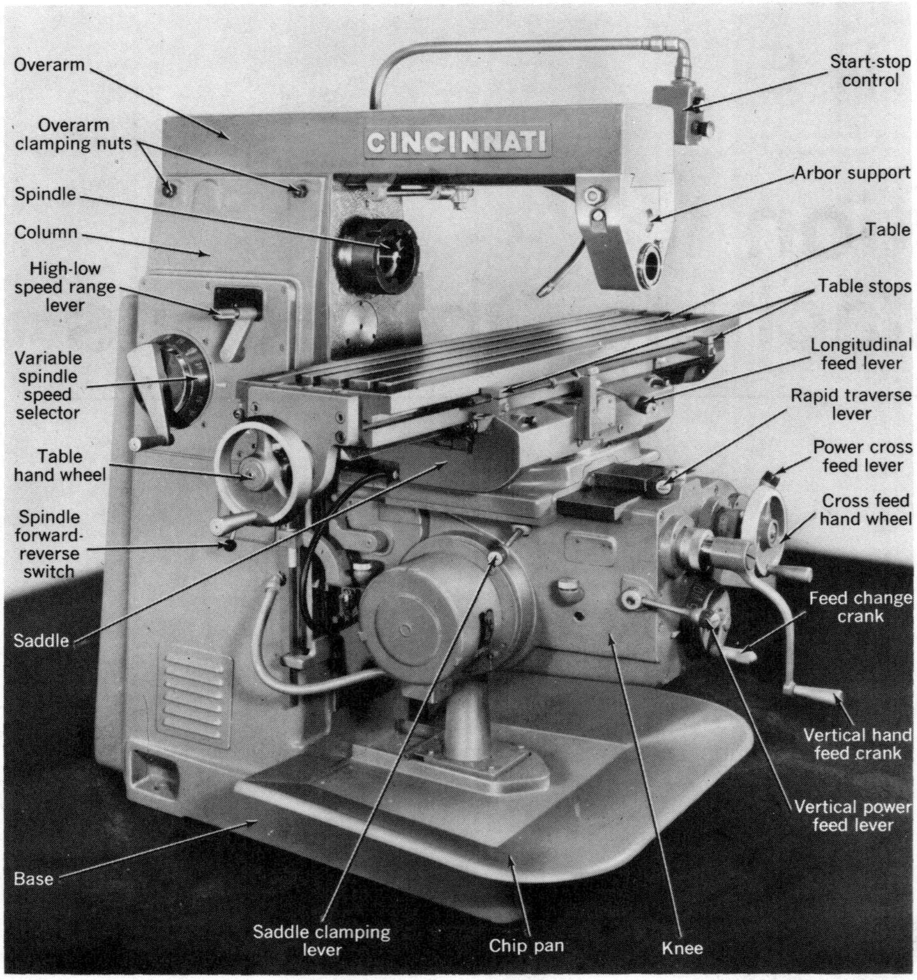

Knee and column horizontal spindle mill

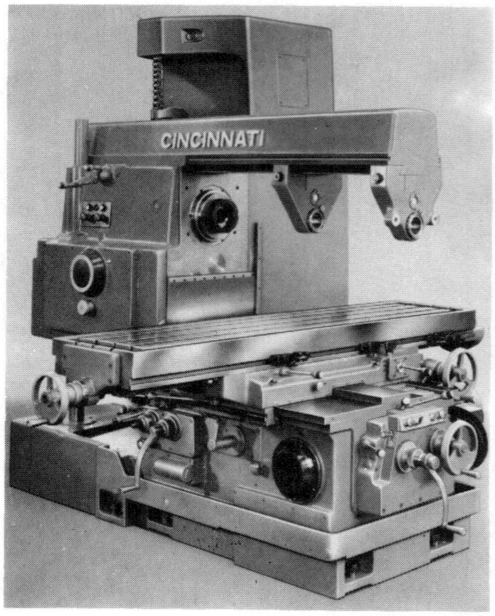

Bed-type manufacturing milling machine

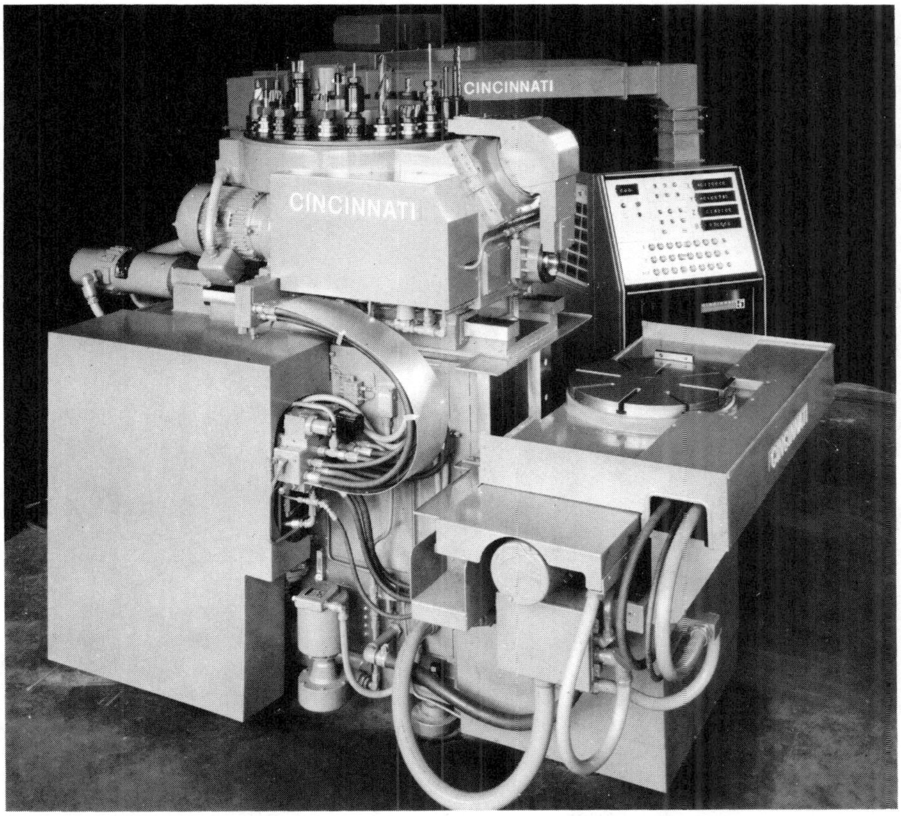

Numerically controlled (N/C) horizontal spindle machining center

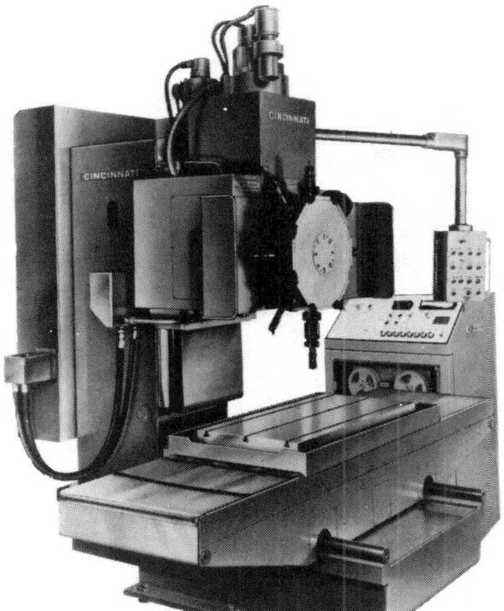

N/C vertical spindle machining center with turret tool changer

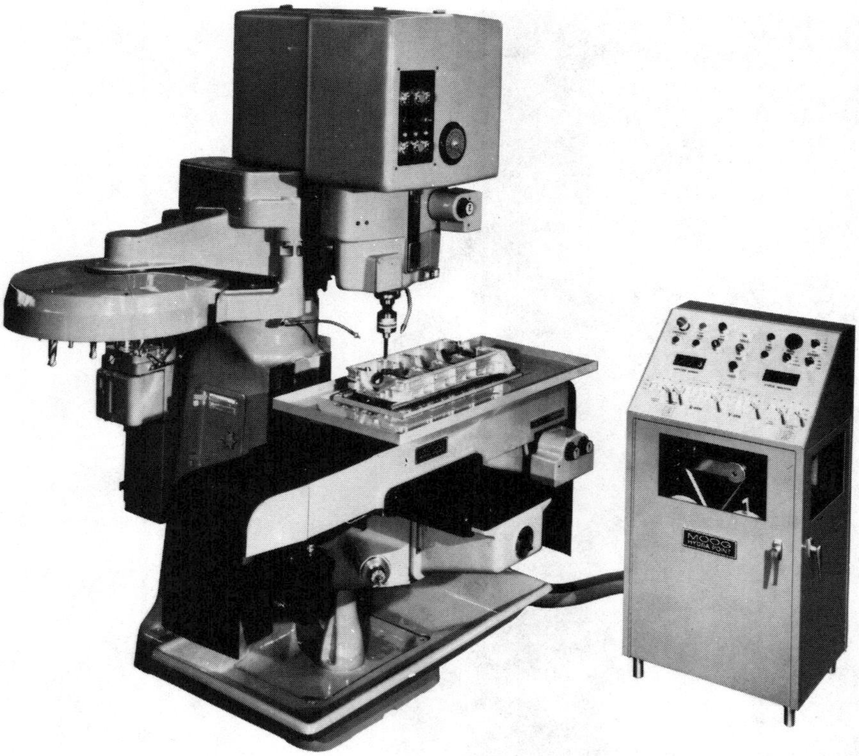

N/C horizontal spindle machining center with carousel tool changer

Vertical multispindle profiler

Horizontal multispindle profiler

Milling machine capability

The primary function of a milling machine is the large group of machining operations called **milling.** In milling, a circular rotating cutter with one or more teeth is in direct contact with the workpiece. The cutter teeth are shaped such that each tooth shaves off a small chip of workpiece material. Depending on the type of milling machine used and the shape and style of the cutting tool employed, a wide variety of machining operations can be accomplished on this machine tool. Typical milling capability includes the following.

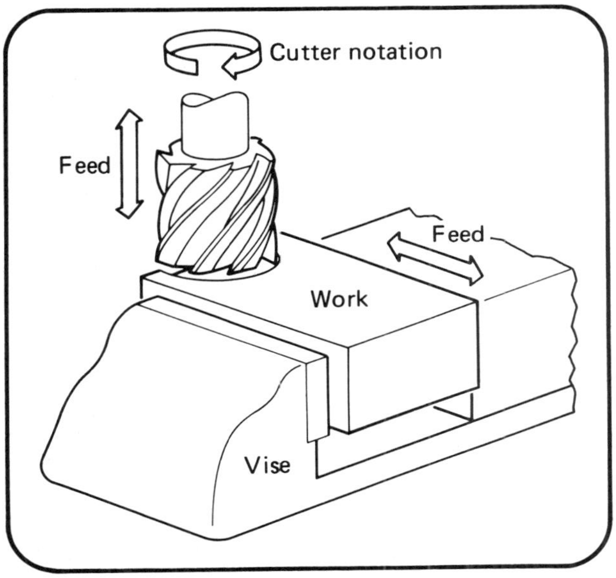

Face or slab mills are used for squaring, that is, bringing a workpiece square on all four sides.

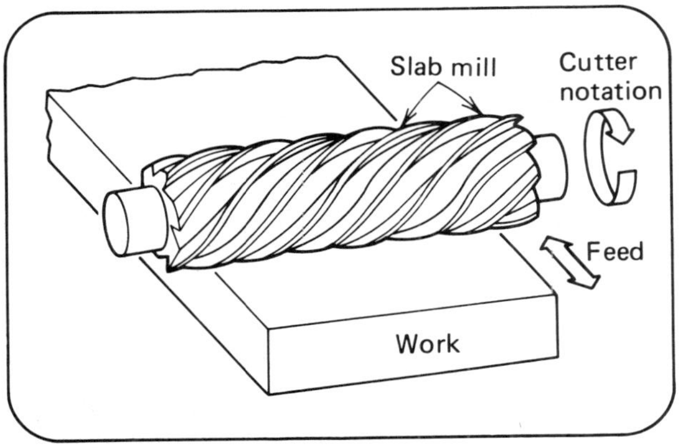

A slab mill may be used to machine large surfaces flat.

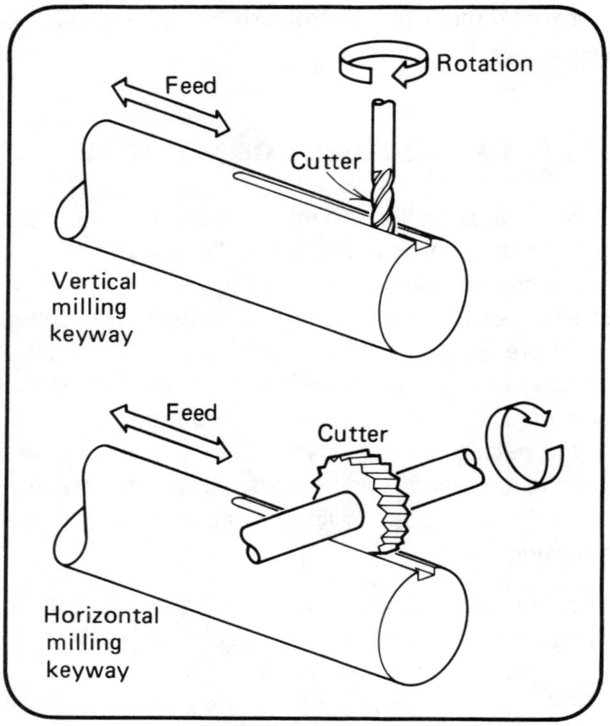

Keyways may be milled using an end mill or milling saw.

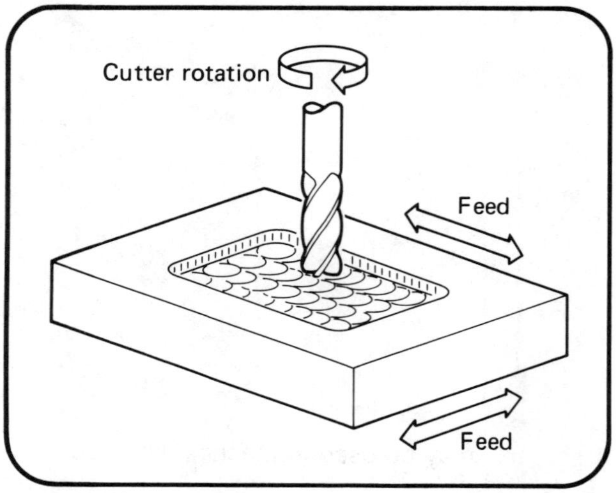

Enclosed pockets are milled using an end mill.

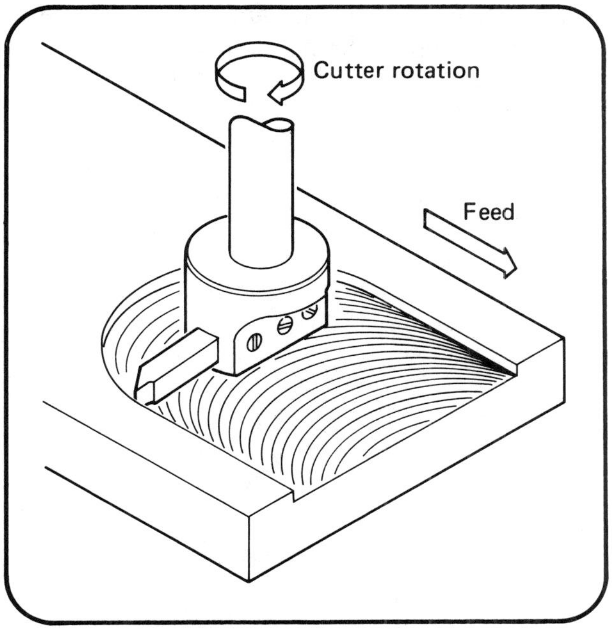

The flycutter uses a single-point tool and is useful for surfacing large areas by taking light cuts, often at high speeds for good surface finishes.

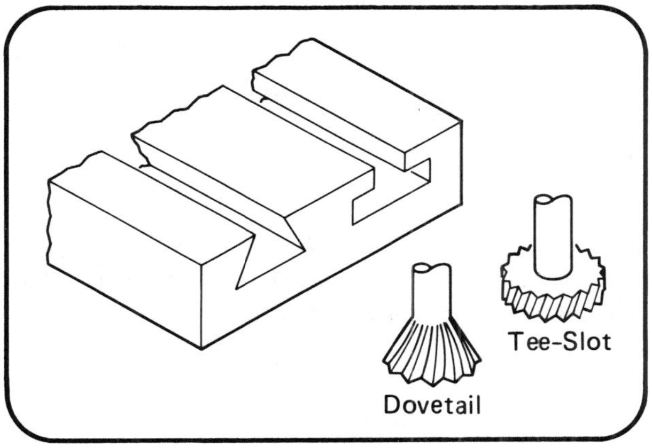

Dovetail and T slots may be cut with the appropriately shaped cutter.

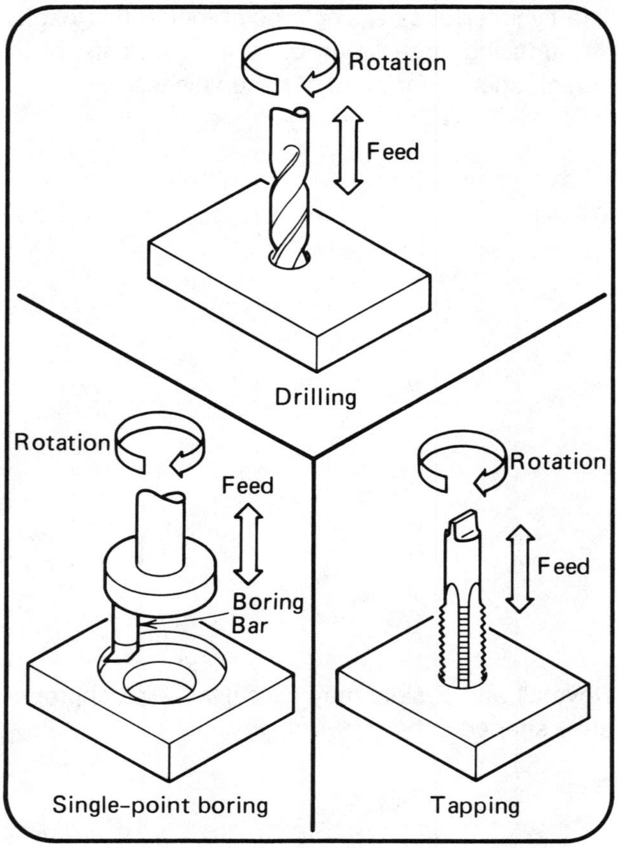

Drilling, single-point boring, and tapping may also be accomplished on the milling machine.

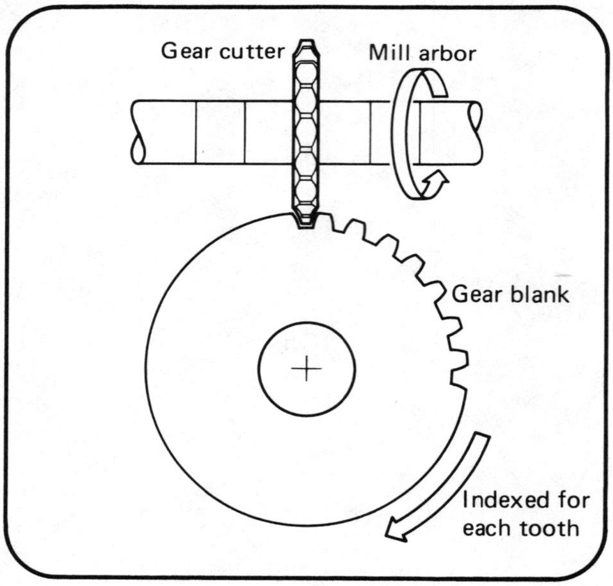

Gears may be cut on horizontal mills using appropriate gear cutters and indexing equipment to provide the required number of gear tooth positions.

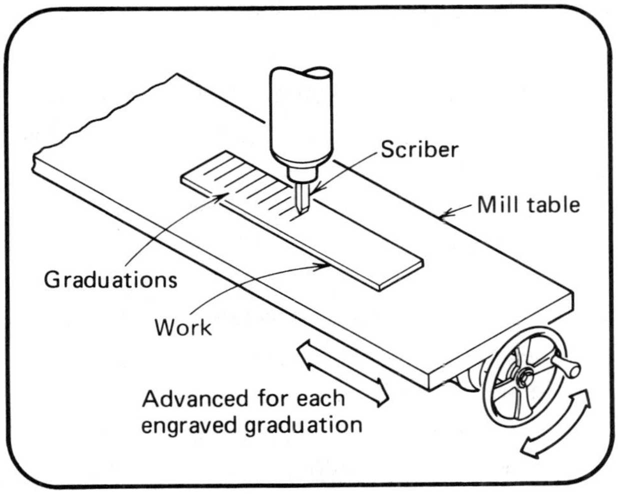

The accurate screw of the mill table may be used to do engraved graduating.

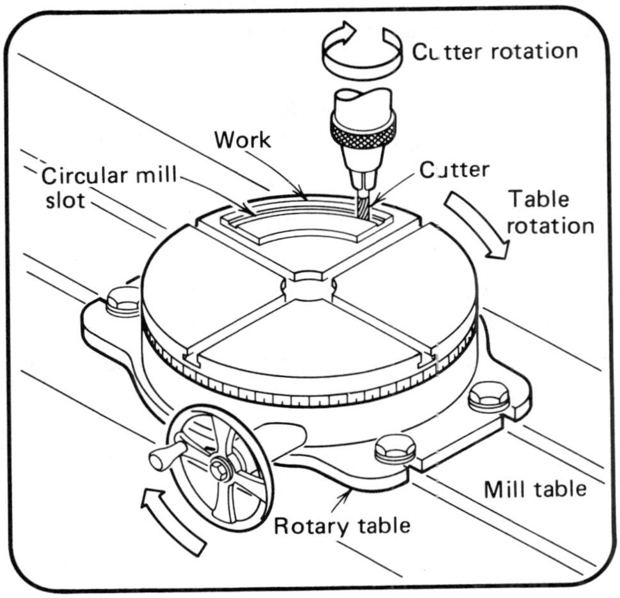

The rotary table secured to the mill table imparts a rotating motion to the workpiece. Many kinds of circular geometry may be machined using this accessory. Hole positions on circles of any given pitch circle may also be drilled or bored using the rotary table.

Milling machine safety

Before operating the machine, review and familiarize yourself with the following safety considerations.

1 Always wear eye protection when in the shop or operating the machine.

2 Learn the location and purpose of all the machine controls.

3 Learn where the motor off/on switch is and how to use it.

4 Be sure to use a hoist and slings when lifting heavy vises and other mill accessories.

5 Use correct feeds and speeds for all machining operations.

6 Do not handle chips with bare hands. Use a brush or rake to remove chips from the machine.

7 Do not attempt to measure the workpiece while it is near a revolving cutter.

8 Beware of any open moving mechanism on the machine that may catch hair or clothing.

9 Handle milling cutters carefully; they are very sharp.

10 Do not allow cutter or arbors to fall onto the mill table after the drawbolt has been released.

11 Be sure that the workpiece is securely and properly clamped.

12 Be sure that cutters and arbors are securely clamped in collets, adapters, and spindles.

13 Listen and feel for excess vibration and noise during a cut. If these symptoms are detected, stop the machine and check the setup.

14 Never leave a running machine unattended.

15 When moving the table, saddle, and knee about by rapid traverse, do not run the workpiece or tooling into the arbor or cutter.

16 Be sure that all locks are tight before beginning a cut. Secure all locks except the component that is moving during the cut.

17 Use correct size wrenches for lock bolts.

18 Keep the floor area around the machine clear of chips, and clean from spills of oil and coolant.

19 Turn the machine off when making any major or electrical repairs.

20 Develop a safety attitude in everything that you do in the machine shop.

Milling machine care and maintenance

For ease of operation, preservation of machine accuracy, and long-term preservation of the machine tool, familiarize yourself with the following care and maintenance factors.

1 Keep the machine clean and properly lubricated.

2 Use the right lubricating oil on the dovetail slides and other sliding surfaces.

3 Use grease in all grease fittings, but do not apply grease to external sliding surfaces, where oil should be used.

4 Do not overtighten locking bolts and clamps.

5 Place tools on tool boards or on the bench, not on the precision surfaces of the machine.

6 When making adjustments and tightening locks, always use wrenches of the correct type and size.

7 Be sure that the spindle is fully stopped before changing speed gears. Do not shift gears while the spindle is turning.

8 Avoid using locks for wear compensation: this will squeeze lubricant from between sliding surfaces, and rapid wear will result.

9 Keep a constant check on oil and hydraulic fluid reservoirs.

10 Be sure that the machine is properly leveled and rechecked from time to time.

11 Never abuse the machine by hammering or forcing any mechanism.

Review questions

1 Name two common milling machines.

2 Name several machining operations that can be done on mills.

3 Point out to your instructor the major points of mill safety.

4 Describe the major points of mill care and maintenance.

5 Point out and name the major parts of the milling machine.

UNIT TWO

Machinability of Materials

Low carbon steel

Low carbon steel is the most frequently used free cutting material. High speed steel or carbide cutters are normally used. A soluble oil and water coolant should be used whenever possible. Chips from face milling should be bright and should curl slightly. They should be thrown clear of the cutter. Slab or end mills should produce long rolled chips. Smaller discolored chips indicate cutter wear.

Alloy steel

Alloy steel is tougher than mild steel. Plenty of coolant should be used. Chips will be bright, but smaller and fairly straight. Cutter wear is indicated by a glazed surface on the workpiece. Use of a worn cutter will form a tough skin on the workpiece (work hardening).

Cast iron

Cast iron has a hard skin or crust from the casting process, and the first cut should be deep enough to cut under this skin. Carbide face mills are normally used. The material removed will be in the form of black crumbly chips. Light cuts will produce black dust. Cast iron should be machined dry and surfaces wiped clean before measuring.

Brass

Brass is a very free cutting material. Chips are removed in a fine spray. Brass generally is machined dry. Cast brass is slightly darker, with chips that are crumbly and slightly curled.

Phosphor bronze

Phosphor bronze is a free cutting but very tough material. Chips are small and spray off the workpiece during face milling. Helical side, face, and end mills with coarse teeth give the best results. A glazed finish on the workpiece indicates cutter wear, and burrs that are formed will be difficult to remove. The use of soluble oil coolant will prolong the life of cutters. Use caution when reaming holes, as the reamer will tend to seize up during the cut.

Aluminum

Aluminum is a very free cutting material. Be careful when clamping the workpiece, as denting or distortion will occur. Coarse tooth cutters will perform the best by permitting good chip clearance. Small chips will be produced by heavy cuts and a fine spray will come from light cuts. Aluminum may be machined dry, but using coolants often will improve surface finish characteristics.

Magnesium alloys

Magnesium is another very free machining material. However, it is very flammable and small chips or powder may ignite. The metal will burn vigorously. Ordinarily there is little danger of fire, as the percentage of magnesium in a given alloy is very small. Chips that are produced will be in powder form or small particles, depending on the depth of the cut.

Plastics

Most plastics are very free cutting and may be machined with a high feed rate and high RPM. This combination will throw the chips away from the cutter, thus preventing the rubbing action that causes dulling. For deep slots, binding may occur if cutters have no side relief. Drills will tend to drill slightly undersized holes. Plastics are usually machined dry.

Review question

Discuss the machinability characteristics of common materials.

UNIT THREE

Planning

How to plan

Good planning is ESSENTIAL for the safe, accurate, and efficient production of workpieces. The chart below is a general guide to the sequence of planning for most workpieces.

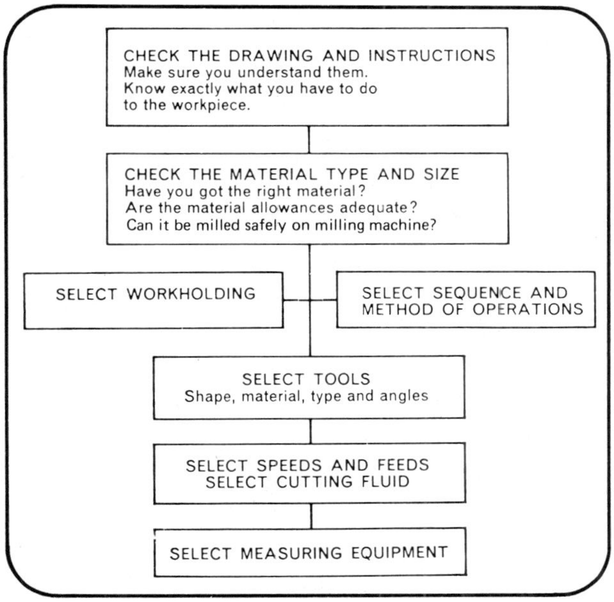

The diagram illustrates the many factors that should influence your choice of the BEST way to produce a workpiece.

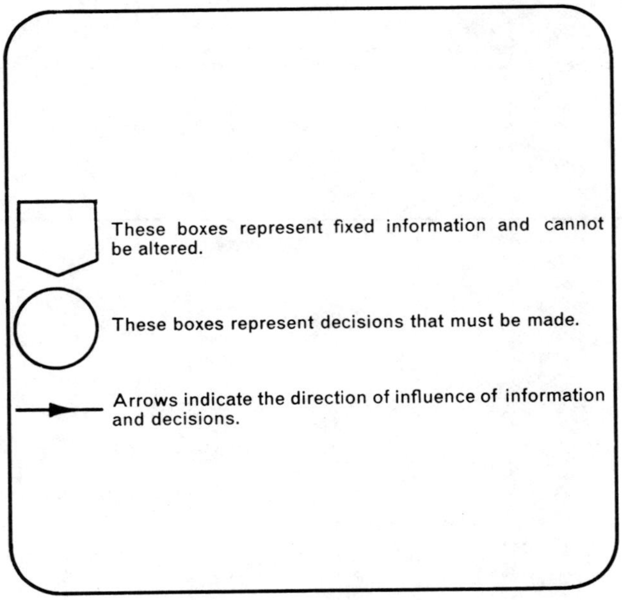

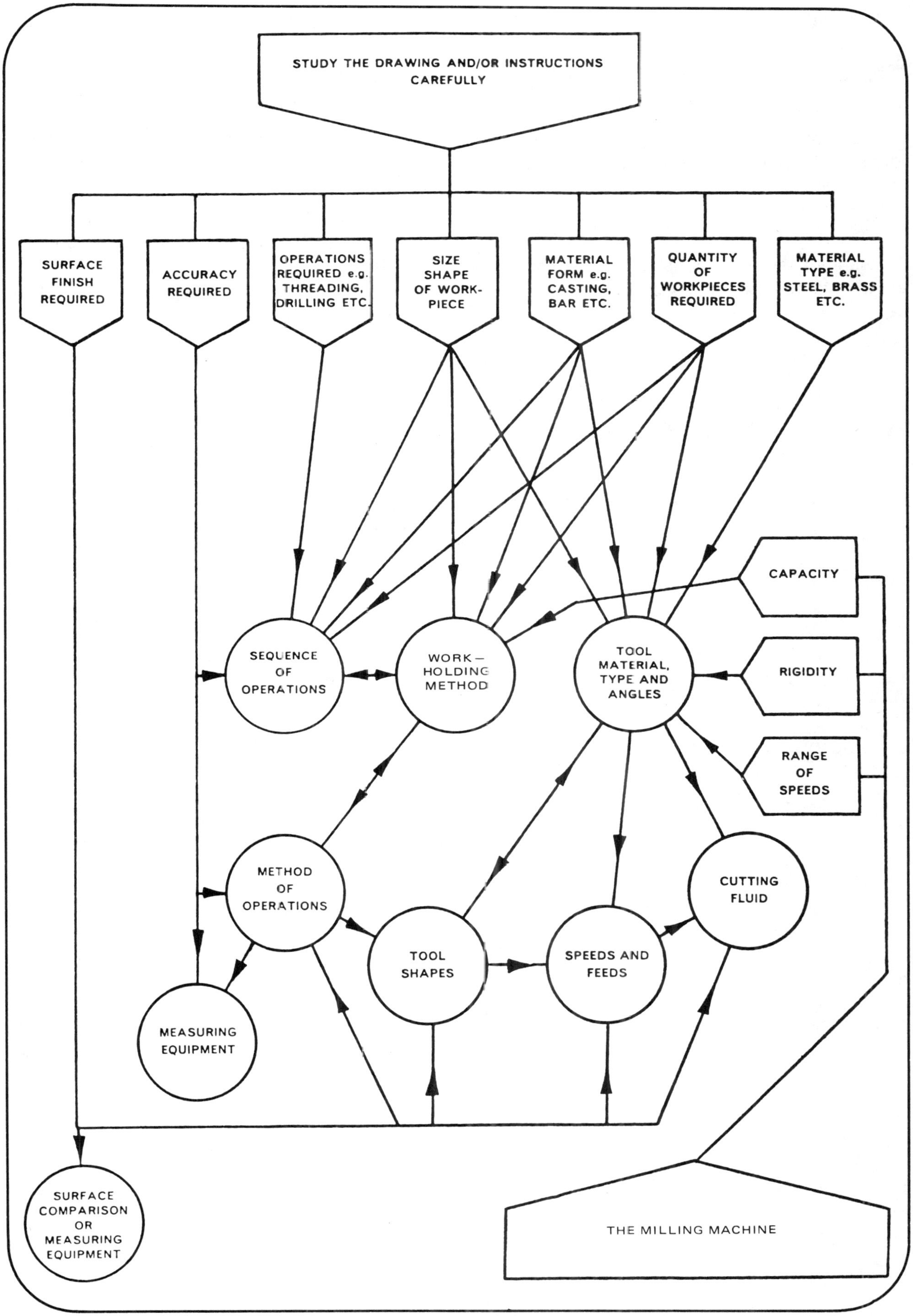

STUDY THE DRAWING AND/OR INSTRUCTIONS CAREFULLY
SURFACE FINISH REQUIRED
ACCURACY REQUIRED
OPERATIONS REQUIRED e.g. THREADING, DRILLING ETC.
SIZE SHAPE OF WORK-PIECE
MATERIAL FORM e.g. CASTING, BAR ETC.
QUANTITY OF WORKPIECES REQUIRED
MATERIAL TYPE e.g. STEEL, BRASS ETC.
CAPACITY
SEQUENCE OF OPERATIONS
WORK-HOLDING METHOD
TOOL MATERIAL, TYPE AND ANGLES
RIGIDITY
RANGE OF SPEEDS
METHOD OF OPERATIONS
CUTTING FLUID
TOOL SHAPES
SPEEDS AND FEEDS
MEASURING EQUIPMENT
SURFACE COMPARISON OR MEASURING EQUIPMENT
THE MILLING MACHINE

Tips on planning the job

1 Study the drawing and job. Ensure that you understand them and know exactly what has to be done to the workpiece. For example, is any excess material to be left on sizes for grinding or hardening?

2 Check material or casting for machining allowance.

3 Plan out sequence of operation. Determine the answers to the following questions.
(a) Can any operations be combined?
(b) Can any operations be eliminated?
(c) Is the sequence the best possible?

4 Select the method of workholding. The simplest is usually the best. Ensure that it will hold the job securely.

5 Select cutters. Base the selection on the operations to be performed, the material and the type of machine.

6 Select measuring equipment. Ensure that it is calibrated.

Review questions

1 Why is planning important to the job?

2 Discuss the important considerations in planning for milling machine work.

UNIT FOUR

Milling Cutters: Types and Applications

Face mills

1 Smaller diameters (usually up to 3 in.), consist of solid, high speed steel cutters mounted on a suitable arbor. These cutters have helical teeth on the periphery, with medium and coarse pitches.

Note:
Cutters up to 6 in. in diameter are available in this type.

2 The larger sizes are made with a body fitted with high speed steel, stellite or tungsten carbide tips, or inserts. Tips may be brazed direct onto the cutter body or brazed onto a holder, which is then mechanically held in the cutter body. Tungsten carbide inserts are mechanically held and can be indexed so that a new cutting edge is presented to the workpiece. When all edges have been used, the insert is discarded or resharpened. These are known as throwaway tips. It is good practice to replace all these inserts in a cutter at the same time.

Note:
These larger size face mills should be mounted directly on the spindle nose.

3 Because of the many variables involved, milling speeds and feeds may have to be varied when all the relevant factors are known.

Types of face mills

Face mills are used to produce flat surfaces.

Typical applications

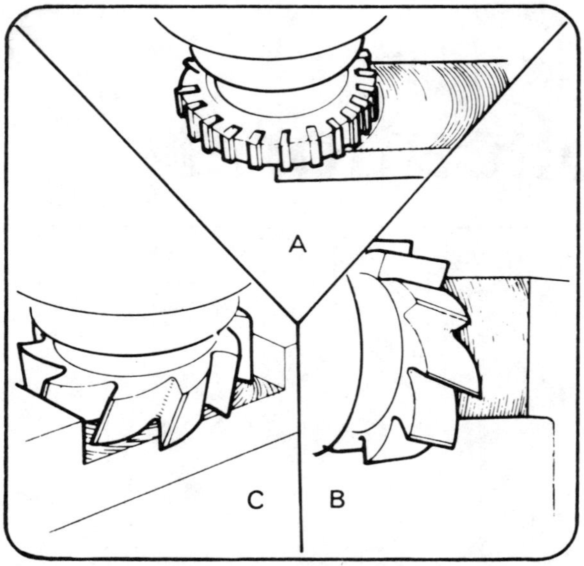

1 Facing the top of a workpiece on a vertical mill (A).

2 Facing the end or edge of a workpiece on a horizontal mill (B).

3 Milling shallow steps on a workpiece (C).

4 Roughing deep steps on a workpiece.

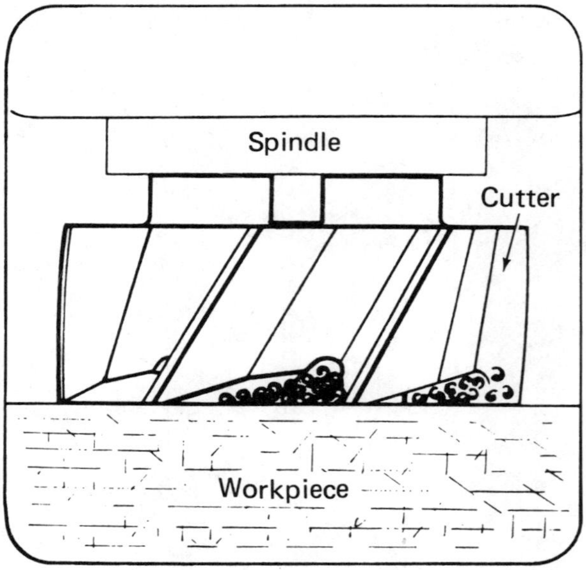

Solid high speed steel cutters

(a) Must have sufficient chip clearance when taking heavy cuts.
(b) Will deal effectively with flat surfaces, steps, and shallow slots and cover most steel alloy ranges, bronzes, and aluminum alloys. Rotate at fast medium speed with medium feed and finish with second light cut.

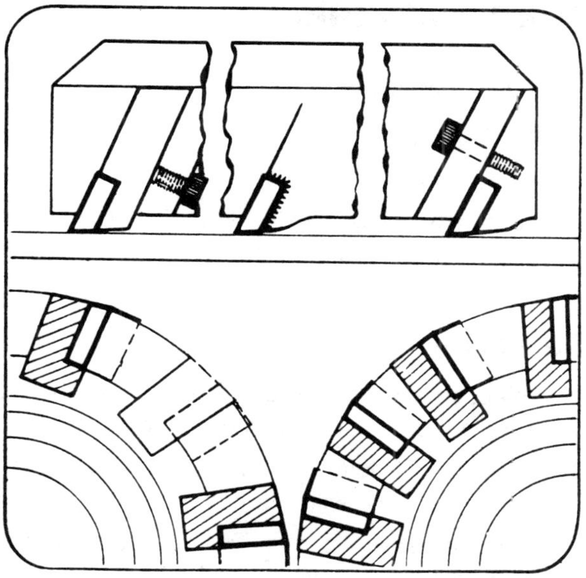

Tipped cutters

(a) Use tipped cutters for larger surfaces and castings. Rotate at medium speed with medium to coarse feed, mainly finishing with one pass. Apply maximum tooth feed to minimize cutter wear.
(b) Remove alternate inserts when using throwaway-type cutters on aluminum and rotate at fast speed.

Note:
Differential spacing of inserts will assist in damping vibrations and eliminating chatter.

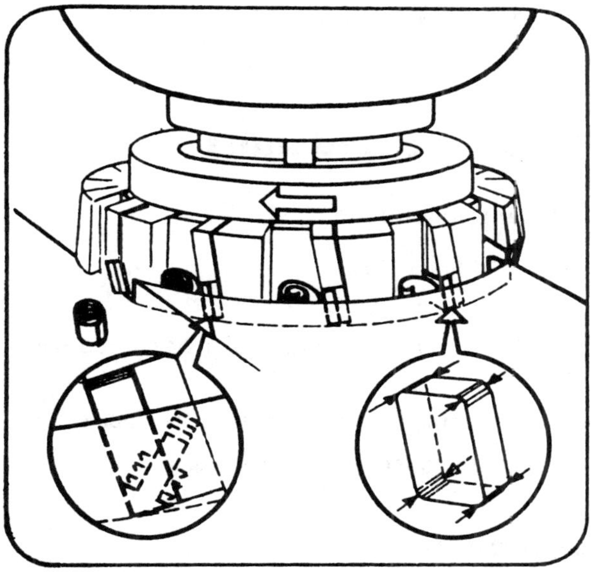

Negative rake

(a) When milling relatively thin workpieces, use cutting force to press workpiece down.
(b) Cutter must have adequate chip clearance because chips do not spiral out of cut.
(c) Ensure that there is no variation of cutter speed by using fast speed and coarse feed, with depth of cut regulated to power available.
(d) Throwaway insert-type cutters offer good economy because several cutting edges can be used.

Selection

1 Select the largest convenient diameter cutter to cover the face to be machined if the face is approximately central to the cutter.

2 Select the smallest diameter cutter to cover the face to be machined, if machining to a shoulder.

3 The number of teeth in the cutter to be used is determined by the material and the rigidity of the workpiece. Coarse-toothed cutters are desirable because they are more efficient, but thin parts require finer toothed cutters to avoid workpiece distortion.

4 The grade of cutter teeth to use is determined by the material and possibly the number of workpieces to be produced. Tipped cutters are best for cast iron and some nonferrous metals.

5 The rake angles (radial and axial) are determined by the material to be machined. As a general rule, the harder the material to be cut, the smaller the rake angles required. Negative rake cutters may be used when a high production rate is required.

Shell end mills

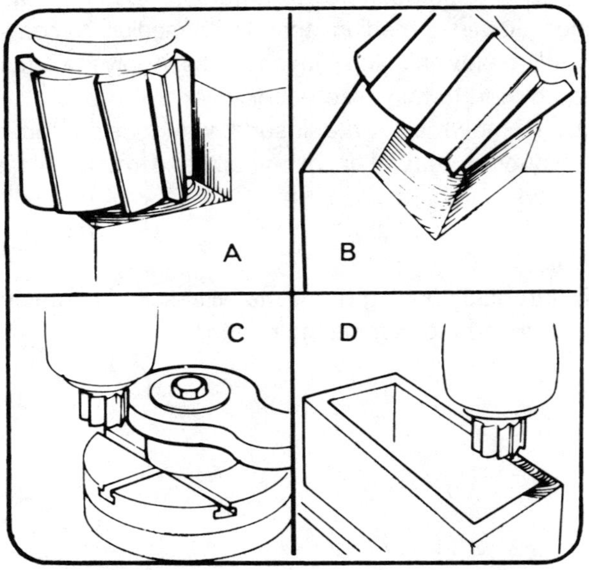

Typical applications

1 Milling a shoulder (A).
2 Milling angular faces (B).
3 Milling internal and external profiles (C).
4 Milling flat faces (D).

Selection

1 Select smallest diameter cutter to cover the face to be machined when milling to a shoulder.

2 Select the largest convenient diameter cutter to machine a flat face with the cutter face.

3 The number of teeth in the cutter to be used is determined by the material and the rigidity of the workpiece. Coarse-toothed cutters with a fast helix are generally more efficient, but thin parts require finer teeth to avoid distortion.

Two-flute end mills

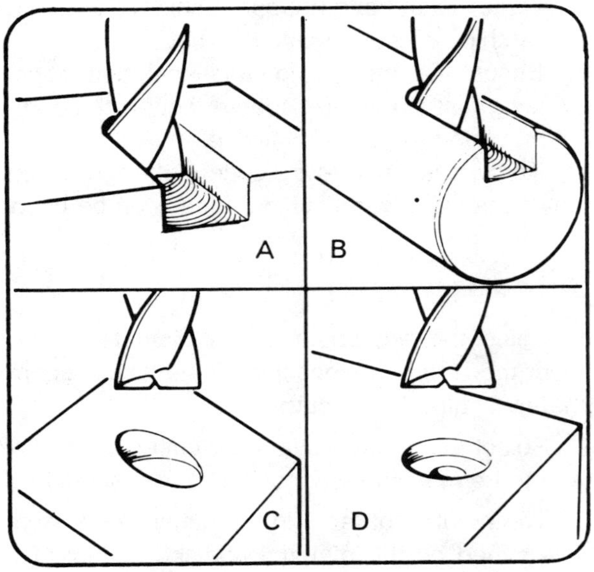

Typical applications

1 Milling slots (A).
2 Milling keyways (B).
3 Milling a flat seat on angular or curved surface (C).
4 Milling small counterbores or spotfacing (D).
5 Realigning incorrectly positioned holes.

Selection

1 The size of the cutter is determined by the width of the slot, or the diameter of the hole to be cut.

Note:
If the slot is not a standard width, the largest standard size cutter under the width of the slot is used, and equal amounts are removed from each sidewall of the slot.

2 The choice of straight or helical teeth is determined by the material and by the rigidity of the workpiece. Helical teeth are desirable for heavy cuts, but thin-walled workpieces or shallow slots are best cut with a straight flute cutter.

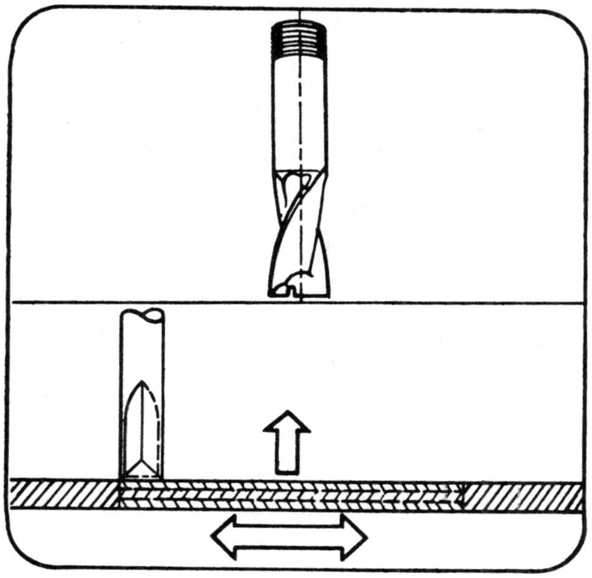

Use spiral fluted end mills center gashed for blind slots. Run at fast speed with coarse feed. Flush coolant to assist chip clearance.

Straight fluted end mills are effective on shallow slots, where chips are easily cleared, and traversing back and forth to reach depth is possible. Use medium speed and feed.

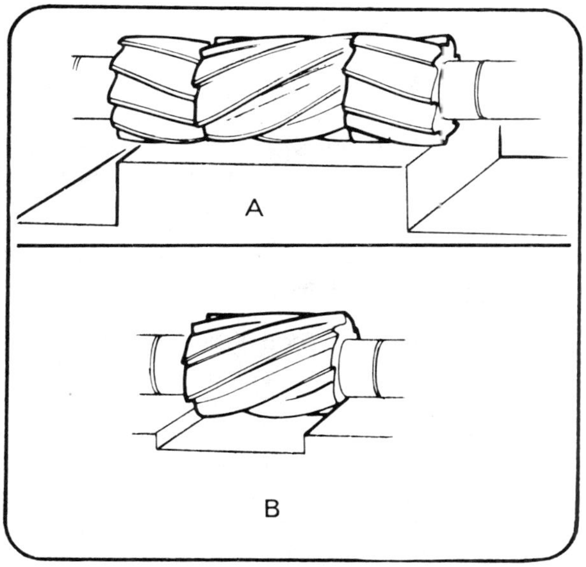

Slab mills

Typical applications

1 Milling the top face of a workpiece (A).
2 Milling a recess in the face of workpiece (B).
3 Milling to a shoulder.

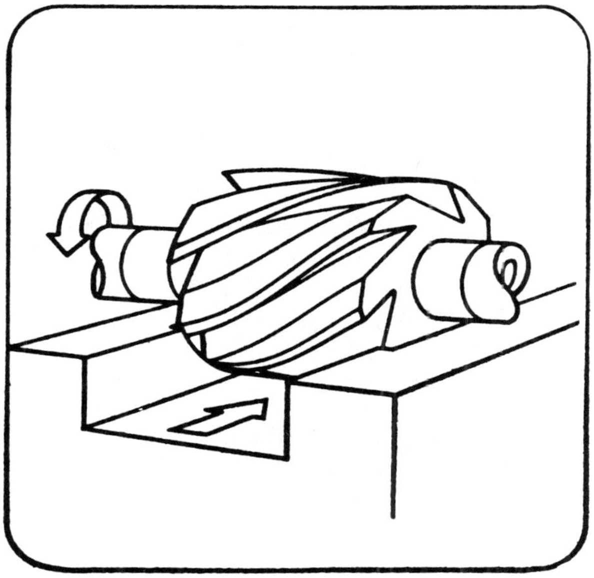

Slab mills are used in many cases for gang milling and require a rigid setup with a slow speed and medium feed on castings and alloy steels.

Light alloys require fast helix and high rake mills, used at medium to high speeds and coarse feeds.

On brittle plastics it is better to climb mill, to prevent breakage of edges on entering and leaving the workpiece. Use low speed and coarse feed.

Pointers on slabmill selection

Select the smallest convenient diameter cutter.

Select a pair of interlocking cutters of exactly the same diameter and with opposite helix angles if a wide face is to be machined.

Side and face cutters

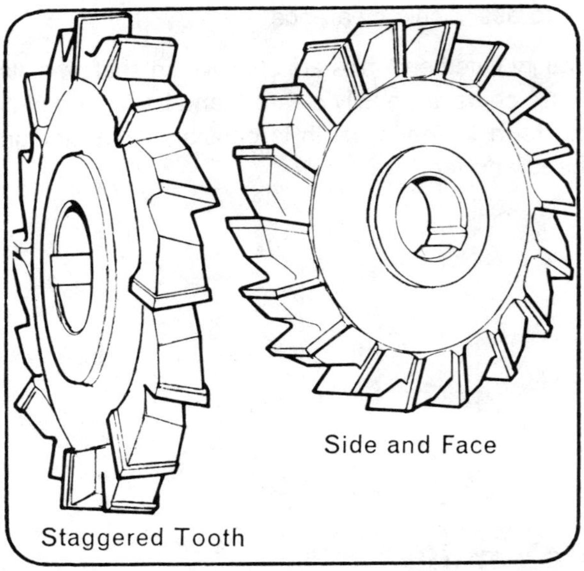

Side and face cutters are disklike cutters with teeth that cut on each side as well as on the periphery. Sometimes a cutter is made to cut on one side only, this type is called a half side milling cutter. Another type has alternate teeth on each side of the cutter; this is a staggered-tooth cutter.

Interlocking cutters are made so that two cutters may be placed side by side with the teeth interlocked. Slots may be cut with these cutters, the width being adjusted by placing shims between the cutter hubs.

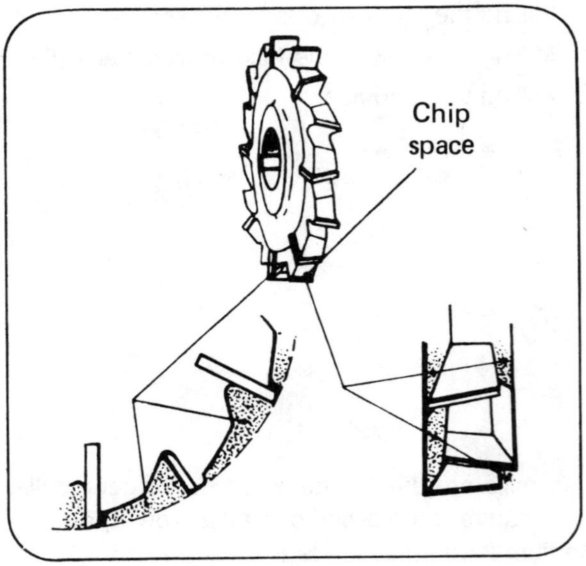

Side and face mills are often used in pairs to mill parallel side faces and are frequently referred to as straddle mills.

When using side and face cutters for deep slots, ensure that there is sufficient clearance for chips.

Use medium speed and feed. When used in a gang, speed is governed by the largest cutter.

Note:
Side and face cutters may be classed as medium and coarse teeth.

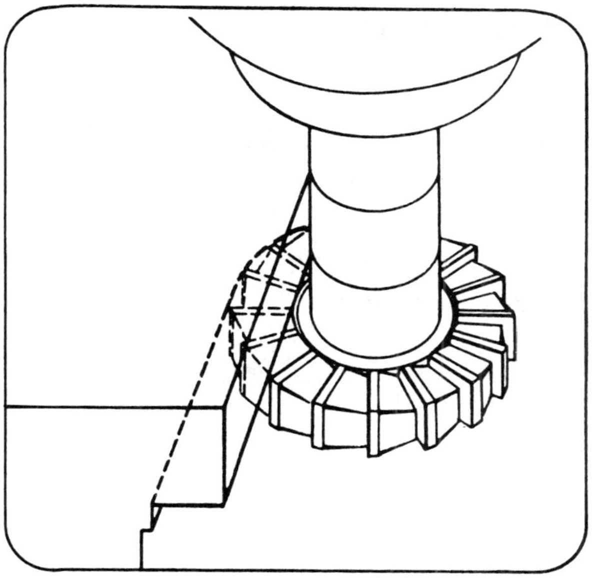

Using a side and face cutter mounted in the vertical head demands medium spaced teeth. To prevent chatter, make sure that more than one tooth is engaged with workpiece during cutting. Use medium speed and fine feed.

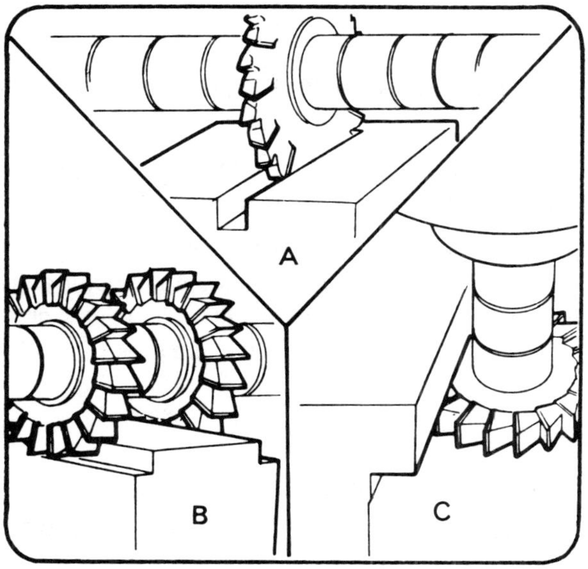

Typical applications

1 Milling a vertical face.

2 Milling a step.

3 Milling a slot (A).

4 Milling 90° vees.

5 In pairs milling parallel faces (straddle milling) (B).

6 Back facing on a vertical mill (C).

Selection

1 Select the smallest convenient diameter cutter.

2 Select a normal type of cutter if only light cuts are required.

3 Select coarse staggered-tooth cutters to take heavy cuts.

4 Select a pair of cutters with opposing angle teeth if setting gangs.

Slitting saws

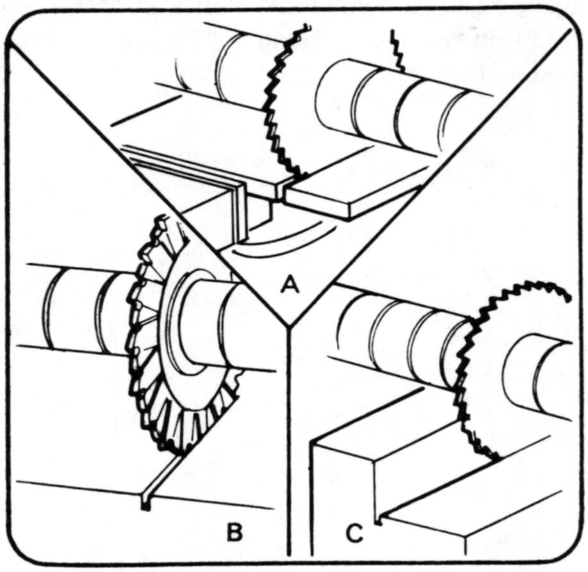

Typical applications

1 Cutting off material to length.

2 Removing excess metal before milling profiles (A).

3 Milling narrow slots (B).

4 Undercutting corners (C).

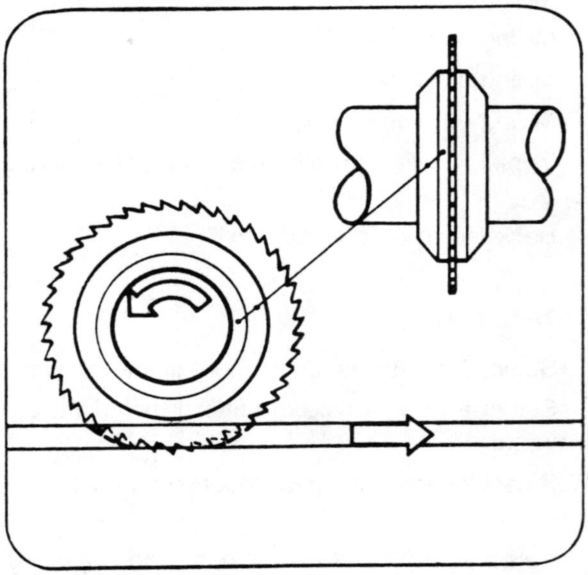

Use slitting saws with fine teeth, fast speed, and fine feed.

There must be support close to cutting edges with minimum allowance to clear workpiece.

Machine a long slot through thin workpieces by sinking through workpiece and traversing to outer edge. This approach, when used in conjunction with climb milling, will resist the tendency of the saw to wander.

Selection

1 Select smallest diameter cutter possible.

2 Select a saw with side teeth for cutting deep and wide slots.

3 Check that teeth corners are not worn or chipped.

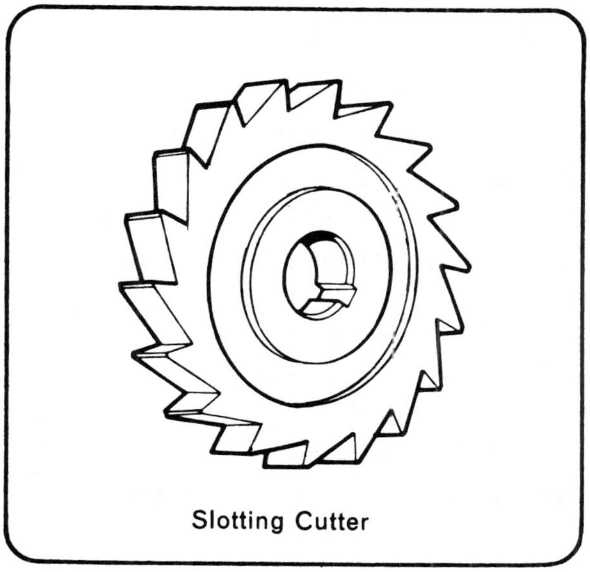
Slotting Cutter

Slotting cutters

Slotting cutters are narrow cylindrical cutters very similar to side and face cutters, but with teeth on the periphery only. The teeth are usually straight on cutters up to ¾ in. (20 mm) wide, whereas wider cutters have helical teeth.

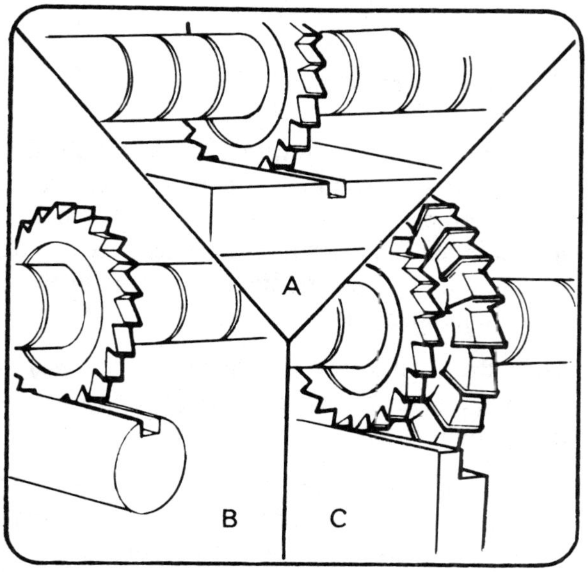

Typical applications

1 Milling slots (A).
2 Milling keyways (B).
3 Milling narrow flat faces (C).

Pointers on cutter selection

Select cutter of correct width when cutting a slot.

Note:
Slotting cutters will not cut on the side, so if a small amount is left on the sidewalls of the slot, it is very difficult to open it up to size.

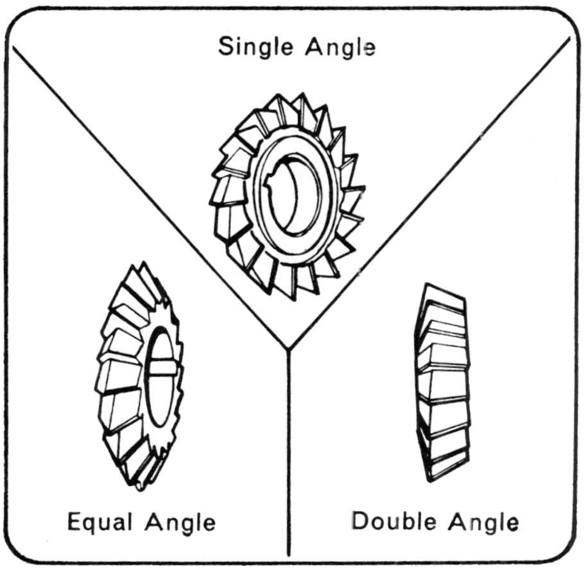

Angle cutters

Angle cutters are similar to small side and face cutters except that one or more sides of the teeth are ground at an angle to the axis.

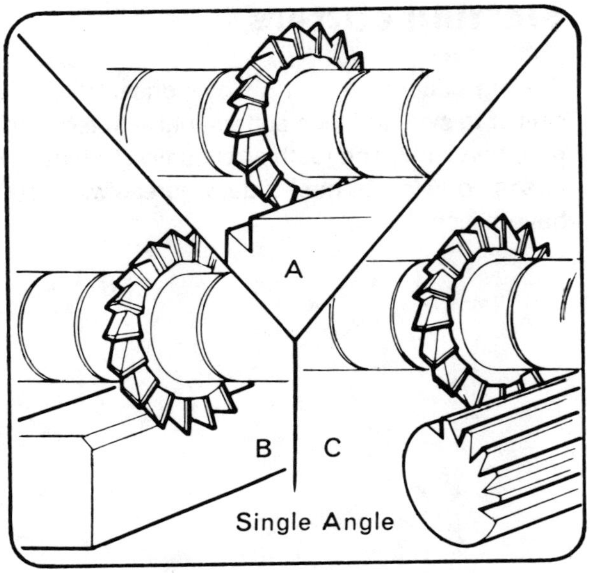

Single Angle

Typical applications

Single-angle cutters

1 Milling an angle on the side of workpiece.
2 Milling dovetails (A).
3 Chamfering the corners of a workpiece (B).
4 Milling straight flutes with radial face (C).

Equal-angle cutters

Milling vee grooves in a workpiece.

Double-angle cutters

Milling helical flutes with radial face.

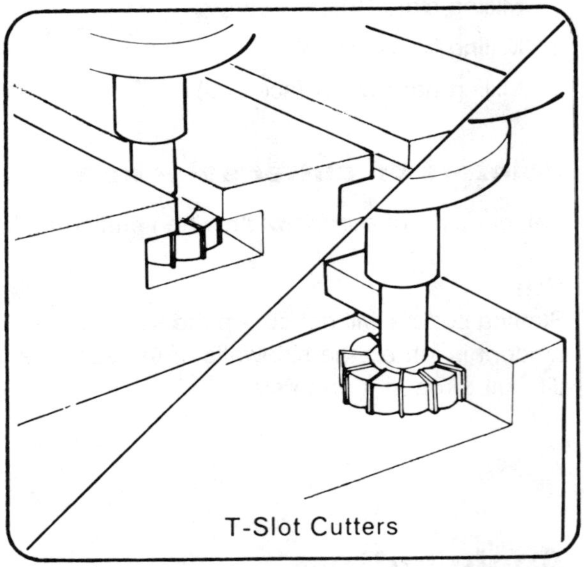
T-Slot Cutters

T-slot cutters

T-slot cutters are made with taper or straight shanks exactly the same as those of end mills and are held in the same way. The teeth are similar to a plain milling cutter but are cut on each side as well as on the periphery. They are often staggered to give better cutting action and more chip clearance.

T-slots are cut in two operations. The primary slot is made with an end mill or slotting cutter. The bottom part is then cut with a T-slot cutter.

T-slot cutters are used almost exclusively to cut T-slots, but they may also be used to cut slots that are inaccessible for other slotting cutters.

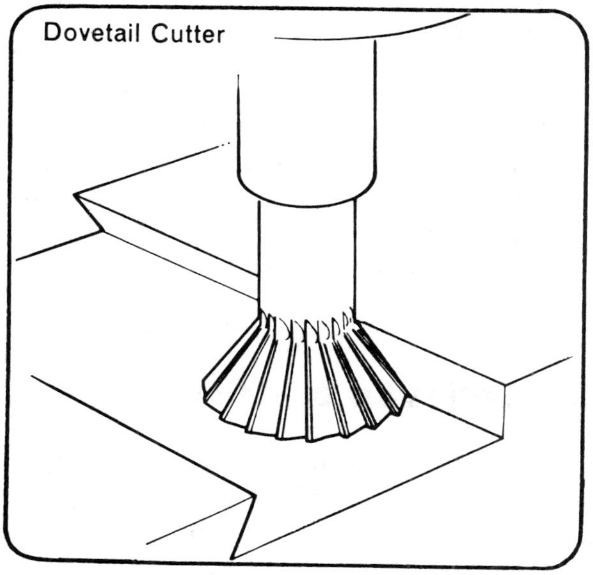

Dovetail cutters

Dovetail cutters are made with taper or straight shanks exactly the same as those of end mills and are held in the same way. The cutting end is similar to a single-angle cutter but is smaller in diameter. The use of this cutter is confined almost exclusively to cutting dovetail slides.

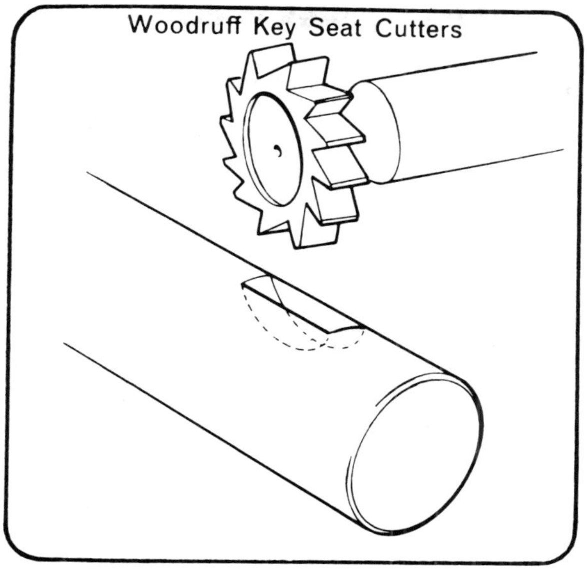

Woodruff key seat cutters

Woodruff key seat cutters are very similar in design to T-slot cutters except that they cut on the periphery only. The teeth may be straight or staggered; the staggered type gives a better cutting action and chip clearance. Woodruff key seat cutters are used to cut curved keyways to suit Woodruff keys. Cutters are numbered according to the size of the keys.

Pointers on Woodruff key seat cutter selection

The number of the cutter required is usually given on the drawing. If instead the diameter width and depth are given, refer to the cutter chart to obtain the correct number for the cutter.

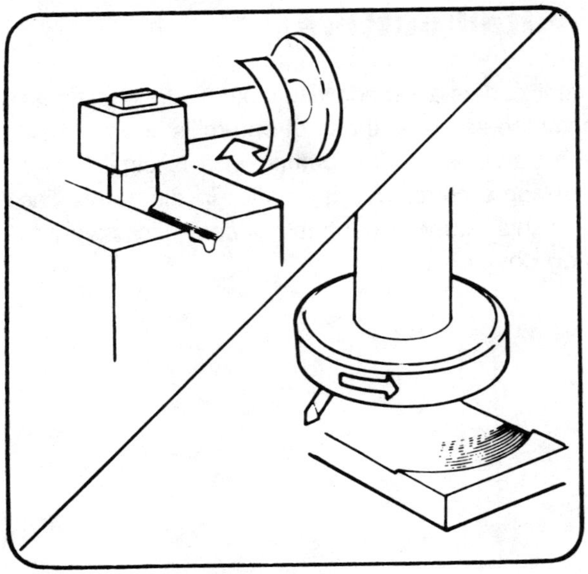

Flycutters

A flycutter is a single-point cutter that is often used to form profiles on short runs of work when it would be uneconomical to make a proper form tool. Flycutters are also used to mill a flat surface on thin plate. The single cutter point almost entirely eliminates any spring of the workpiece.

Form-relieved cutters

Form-relieved cutters are made in many shapes and sizes and play an important part of milling. They are used to duplicate accurately a profile of special shape and should be used only for the purpose for which they are designed.

Types of form-relieved cutter

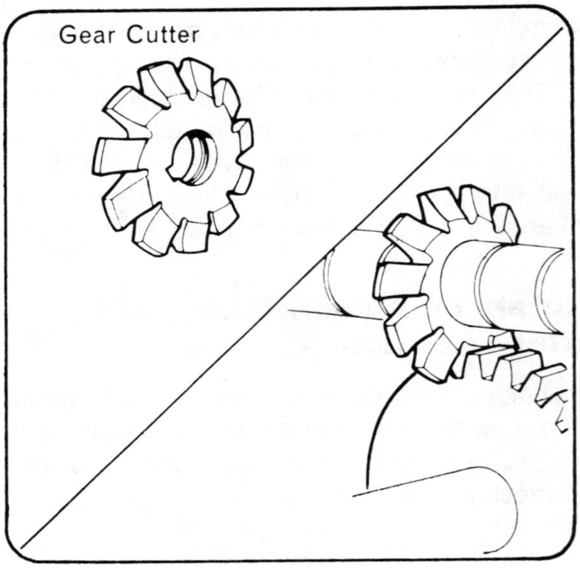

1 *Gear cutters* Used to mill teeth on spur gears.

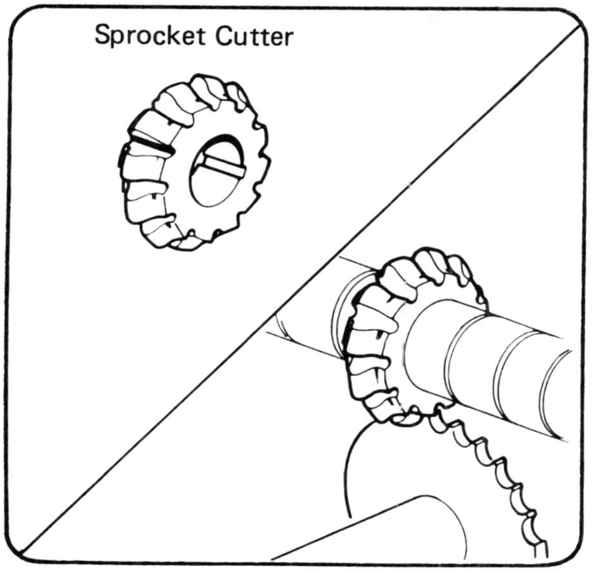

2 *Sprocket cutters* These are sometimes used on milling machines to form the teeth of chain wheels and sprockets. The correct cutter, which is usually stated on the drawing, must be obtained because each cutter is made to cut a specific number of teeth.

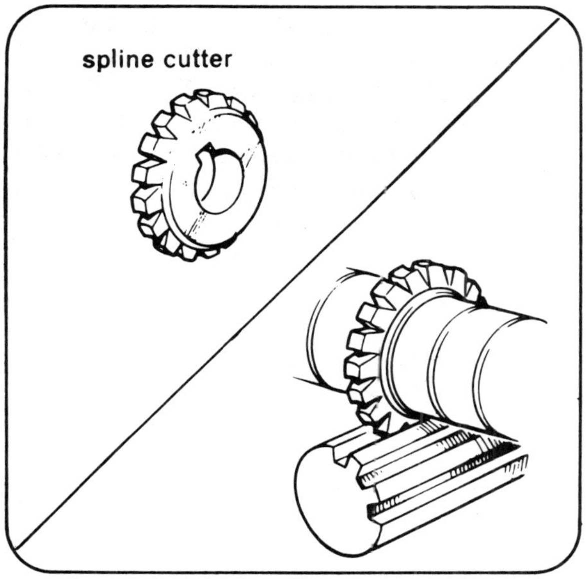

3 *Spline cutters* These are made to form the bottom diameter and the keys of splines in one operation. They are usually made to special requirements.

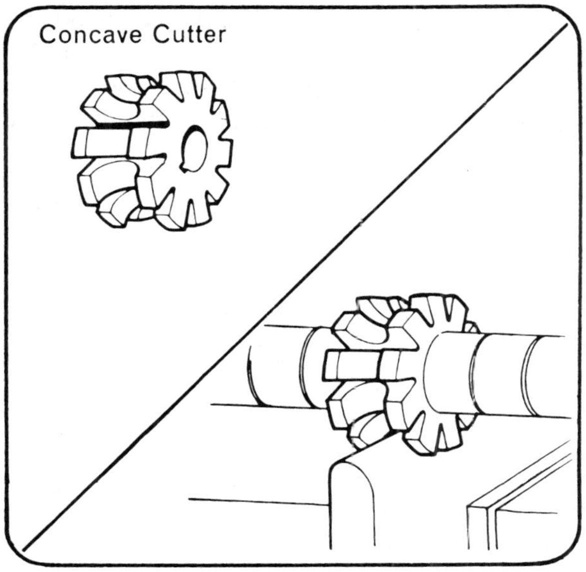

4 *Concave cutter* This cutter is used to form a radius on the face of a workpiece.

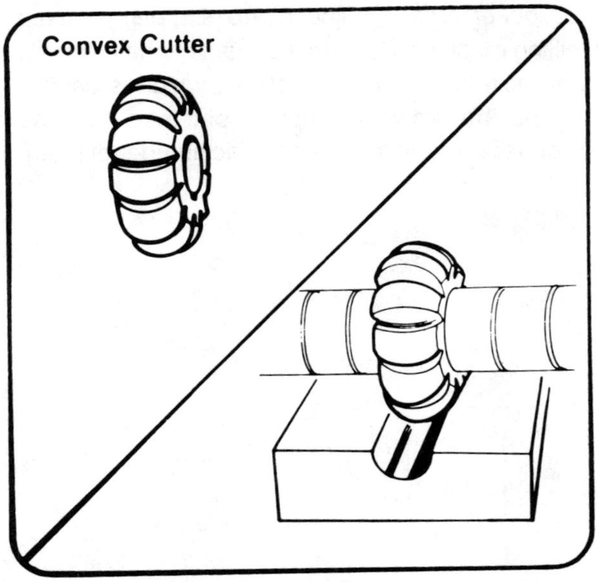

5 *Convex cutter* This cutter is used to form an internal radius or groove on a workpiece.

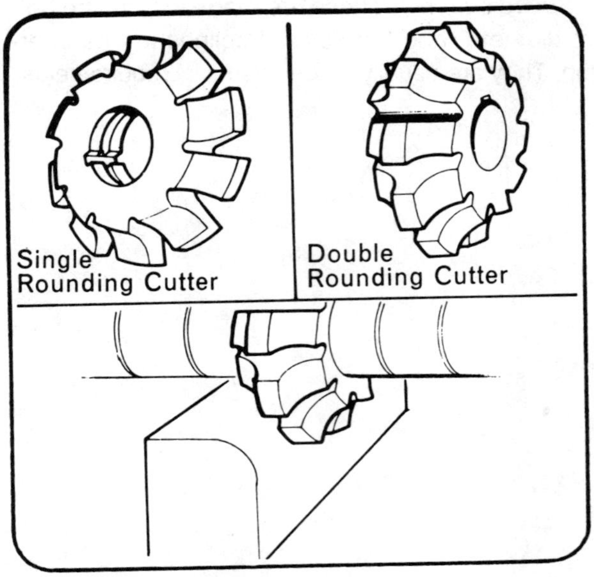

6 *Corner rounding cutters* Such cutters may be single or double sided and are used to form a radius on the corner of a workpiece. The radius of the cutter does not form a full quarter circle, but blends into tangents to the radii of approximately an 80° inclusive angle, to give side and top relief.

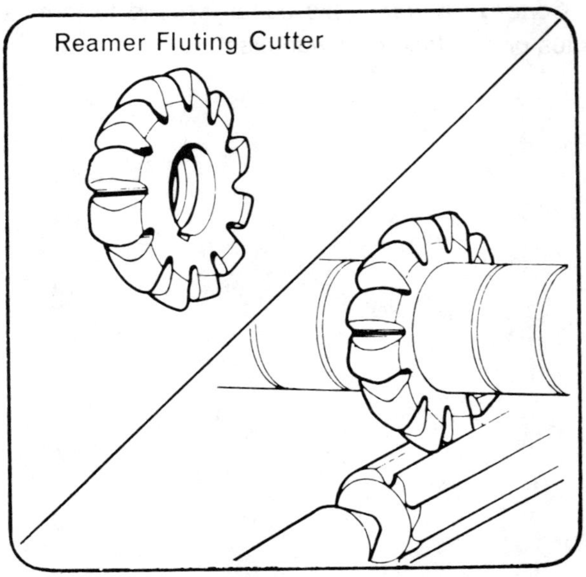

7 *Reamer fluting cutters* These single-purpose cutters are used for forming the flutes of reamers. The cutters are numbered 1 to 9 according to the number of flutes and the diameter of the reamer to be cut.

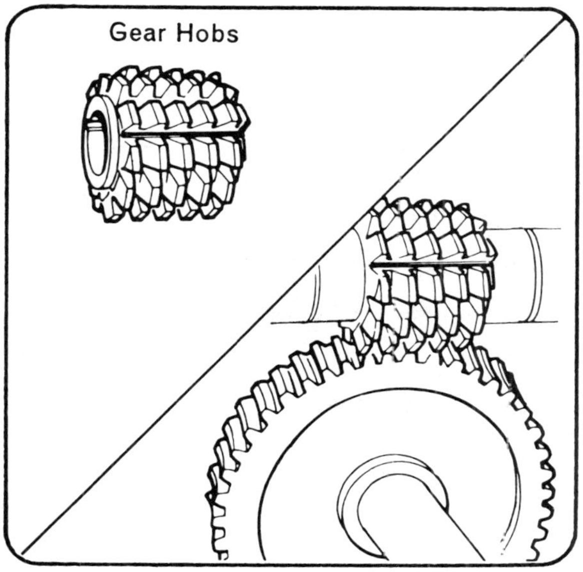

8 *Gear hobs* These are multiple-tooth cutters, and are occasionally used on milling machines to finish worm gears, which have previously been machined to leave a light finishing cut to be taken.

9 *Special form cutters* Cutters having special forms may be made to suit the requirements of a wide variety of workpieces.

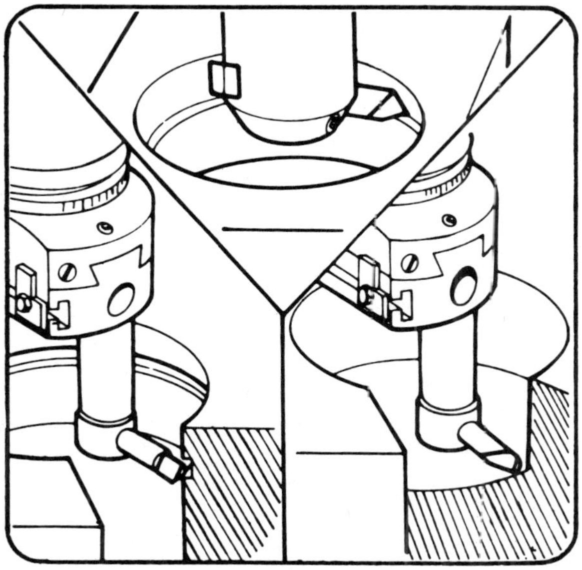

Boring tools

The boring of holes is an operation that is often carried out on a milling machine.

Boring tools may be held in boring bars, which are held directly in the machine spindle, or in a boring head.

Typical applications

1. Boring holes, through or to a given depth.
2. Facing the bottom face of a bore.*
3. Boring a groove or recess in a bore.*
4. Turning an outside diameter.
5. Grooving or recessing a face.*

* Operation applies only to boring heads that may be adjusted while rotating.

Review questions

Describe the general applications of the following milling cutters:

(a) Face mill.

(b) Shell end mill.

(c) Two-flute end mill.

(d) Slab mill.

(e) Side and face cutter.

(f) Slitting saw.

(g) Slotting cutter.

(h) Angle cutter.

(i) T-slot cutter.

(j) Dovetail cutter.

(k) Woodruff key seat cutter.

(l) Flycutter.

(m) Form-relieved cutter.

(n) Boring tools.

UNIT FIVE

Mounting Cutters

Mounting cutters direct to machine spindle: Vertical mill

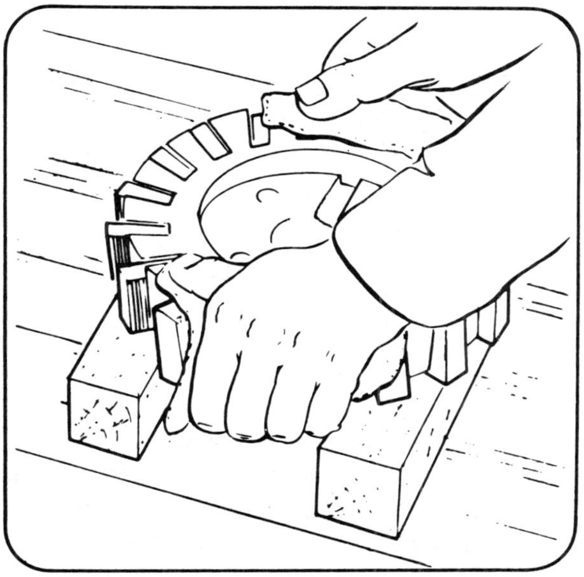

1 Position cutter on table

(a) Place two blocks of wood or soft material on the table, to support cutter.

Note:

Ensure that the maximum possible amount of cutter face will rest on the blocks, but with sufficient space between the blocks to enable the hand to engage the securing screws, through the cutter body into the spindle nose.

(b) Lift cutter, and place teeth down on top of blocks.

(c) Clean face of spindle nose and counterbore of cutter.

Safety

Protect your hands with a rag when lifting the cutter.

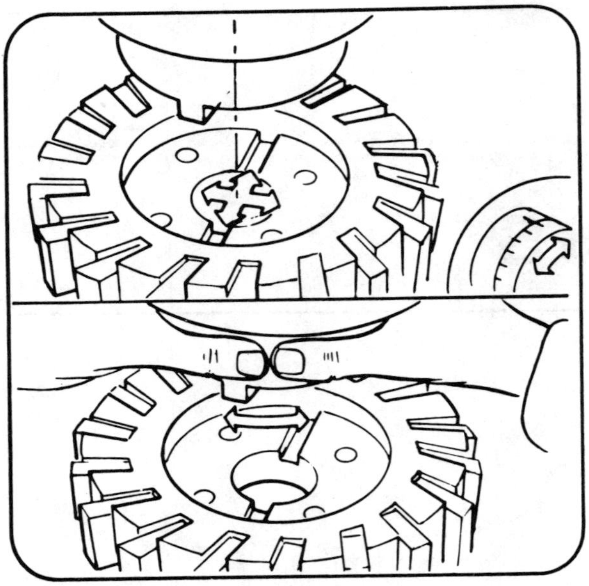

2 Center machine spindle

(a) Raise machine table until top face of cutter is approximately 3 to 4 in. (75 to 100 mm) below spindle nose.

(b) Adjust table position to position diameter of spindle nose central to counterbore of cutter.

(c) Disengage drive to allow spindle to be rotated by hand. Align spindle drive dogs with slots at bottom of counterbore of cutter.

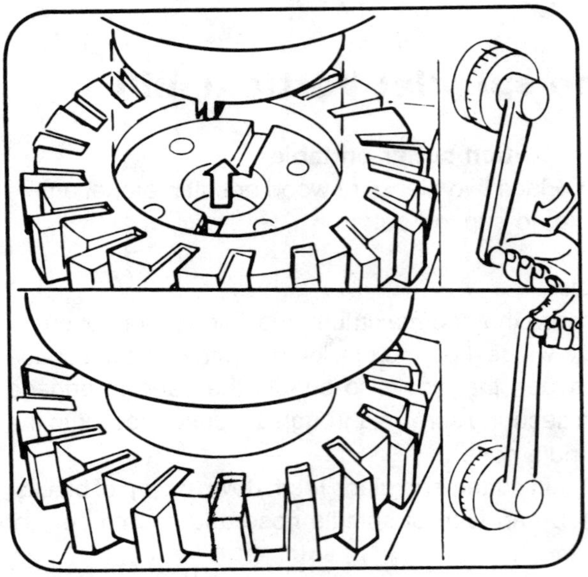

3 Engage cutter to spindle

(a) Hand crank table up, visually checking that spindle nose is centered to cutter counterbore.

(b) Hand crank table up slowly until solid resistance to turn of hand crank is just felt.

Note:
Do not forcibly turn the hand crank when resistance is felt, or the lead screw will be damaged.

4 Secure cutter to spindle nose

(a) Position securing screw (usually socket head screw) through cutter body, and finger tighten, ensuring that head fully enters counterbore.

Note:
If the threaded hole in the spindle nose is not correctly aligned to the hole in the cutter body, the drive dogs are not correctly engaged in the drive slot. Lower the table and realign.

(b) Position other securing screws and finger tighten.

(c) Lower machine table and remove blocks.

(d) Tighten screws securely using correct size key.

Note:
When disengaging the cutter from the machine spindle, use reverse procedure.

Mounting cutters direct to machine spindle: Horizontal mill

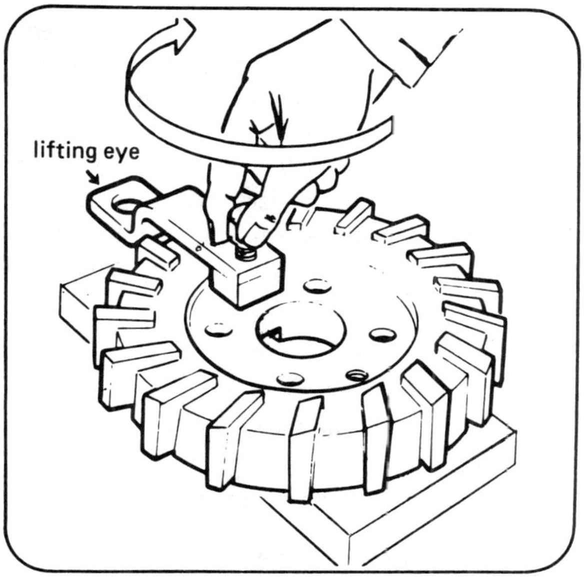

1 Attaching lifting eye

(a) Fasten lifting eye to front face of cutter body, aligning clearance hole to tapped hole in cutter.

Safety

Ensure that an adequate bolt thread is engaged.

(b) Swivel eye to align to cutter centerline.

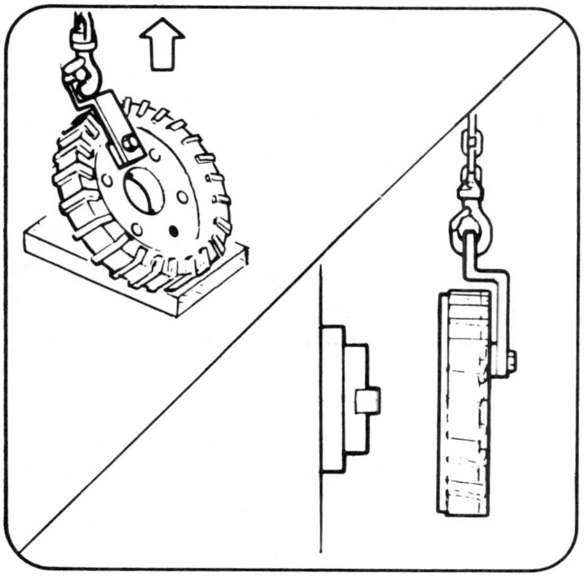

2 Lift cutter

(a) Position crane directly over cutter.

(b) Engage crane hook in eye of lifting eye and raise steadily, allowing cutter to lift to vertical position.

(c) Raise cutter to approximate height of spindle nose.

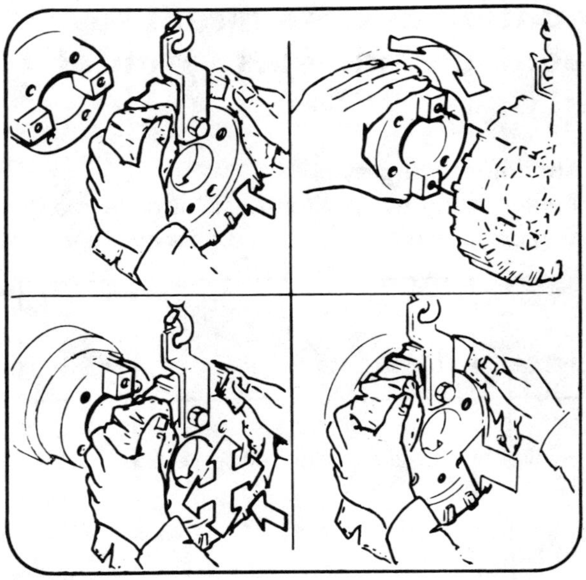

3 Engage cutter on spindle

(a) Lower table to be well clear of cutter.

(b) Clean face of spindle nose and counterbore of cutter.

(c) Push cutter to position counterbore approximately 3 to 4 in. (75 to 100 mm) from spindle nose.

(d) Disengage machine spindle drive and rotate spindle, visually aligning drive keys to slot in cutter counterbore.

(e) Push cutter toward spindle nose, adjusting height with crane until cutter slides easily onto spindle nose.

(f) Rock cutter slightly, attempting to rotate it to ensure that drive keys are engaged, and push cutter in until solid resistance is felt.

> **Safety**
>
> Protect your hands from the sharp edges of the cutter with a rag.

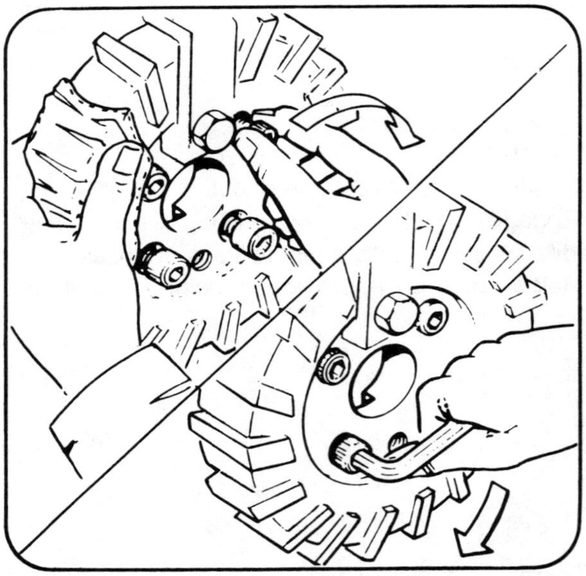

4 Tighten cutter on spindle

(a) Hold cutter in position and engage securing screws in spindle nose.

(b) Tighten securing screws, ensuring that heads fully enter counterbores.

5 Remove lifting eye

(a) Disengage crane hook.

(b) Loosen lifting eye screw and remove.

> **Safety**
>
> Push the crane hook into a safe position clear of aisles and work area.

Note:

If no lifting eye is available, the cutter may be lifted into position manually. Assistance will be required to engage and tighten the securing screws.

Mounting taper shank cutters

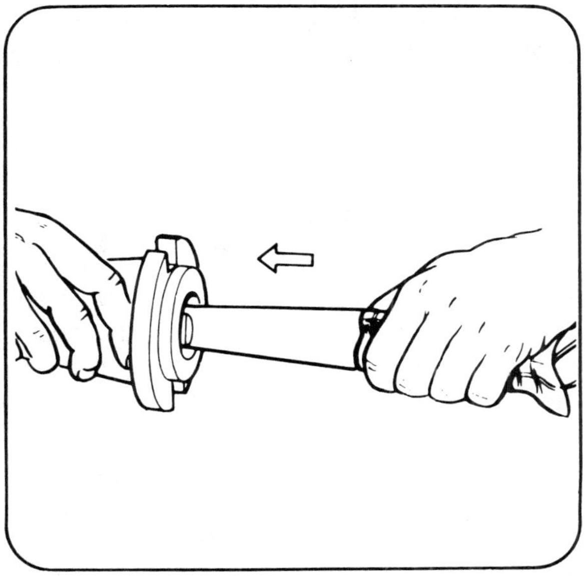

1 Select adapter

(a) Use shortest adapter possible.

(b) Internal taper should be same size as cutter taper if possible.

(c) If necessary, use sleeves, but keep number used to minimum.

2 Engage cutter in adapter

(a) Clean taper in arbor and cutter shank and remove any burrs.

(b) Align cutter tang to slot in taper, and engage tapers with a sharp push.

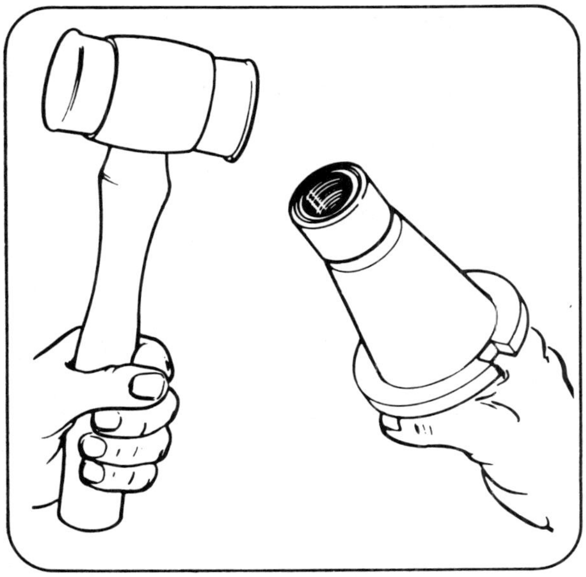

(c) Tap holder in with soft hammer.

Safety

Be careful of sharp edges on cutters.

3 Engage adapter in machine spindle

(a) Clean both tapers.

(b) Engage drive keys in arbor slots.

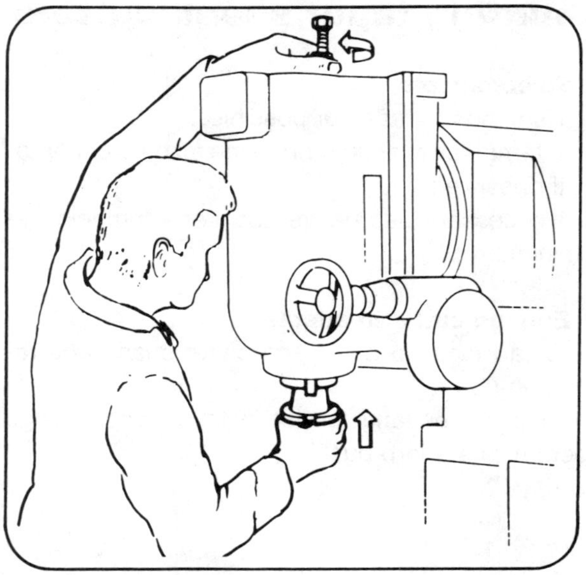

(c) Hold cutter in position applying upward pressure.

(d) Screw drawbolt well into adapter.

(e) Tighten drawbolt locking nut.

(f) Rotate spindle at a moderate speed, to visually check cutter runout.

Note:
When using taper cutters with a tapped shank, an open-ended adapter should be used to allow the drawbolt to screw directly into the cutter tang.

When removing this type of cutter from the adapter, screw a bolt of the same thread size into the tapped end of the shank and tap the head of the bolt with a soft hammer to dislodge the cutter.

Mounting straight shank cutters

Mounting cutters in automatic locking chucks

1 Insert collet arbor in machine spindle

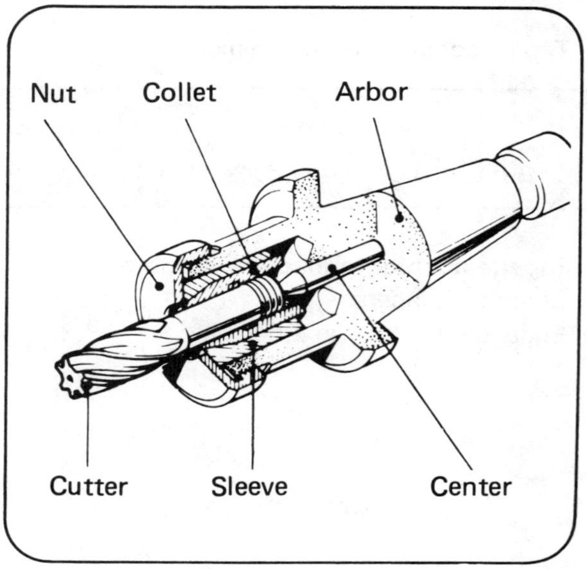

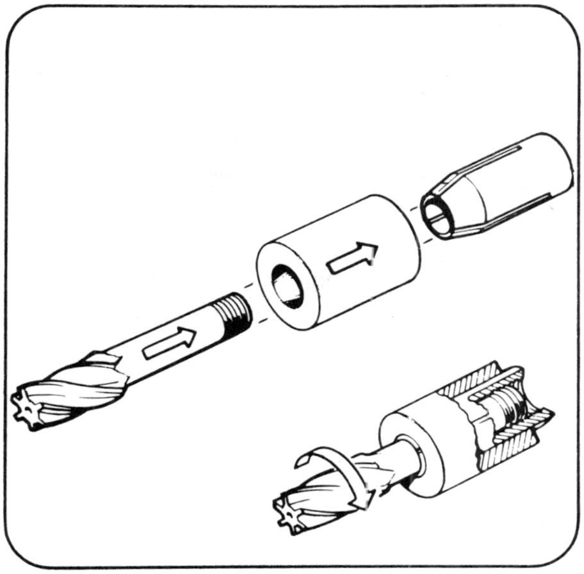

2 Insert collet assembly in arbor

(a) Select appropriate size collet and insert it into locking sleeve.

(b) Insert assembly into bore of body, making sure that driving keys are engaged in the two holes at end of bore.

(c) Screw on retaining nut until it contacts endface of sleeve and unscrew one half turn.

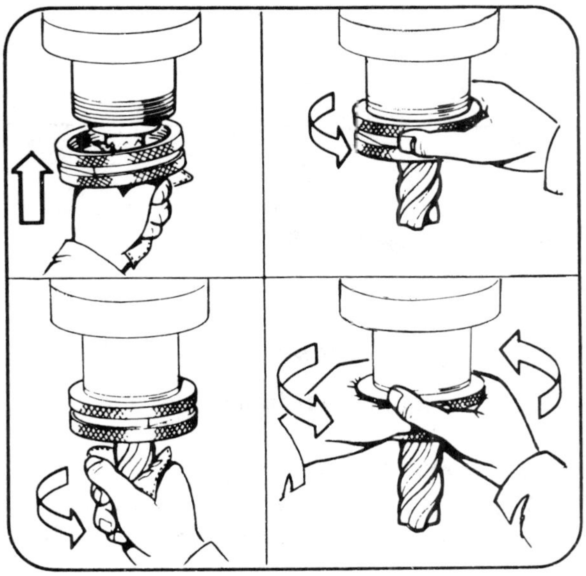

3 Insert cutter in collet

(a) Screw in cutter until it contacts center and becomes tight.

(b) Lightly tighten nut with correct wrench.

(c) Screw nut onto arbor until slight resistance is felt.

(d) Screw in cutter until solid resistance is felt, to engage arbor center in center hole of cutter.

(e) Hand tighten cutter securely.

(f) Hand tighten nut securely or lightly tighten with correct wrench.

Safety

Protect your hands with a rag when screwing the cutter into the collet.

Note:
Cleanliness is essential. Be sure that all parts of the chuck are thoroughly clean before assembly.

Removing cutters from automatic centering chucks

Remove cutter

(a) Engage low gear on machine spindle.

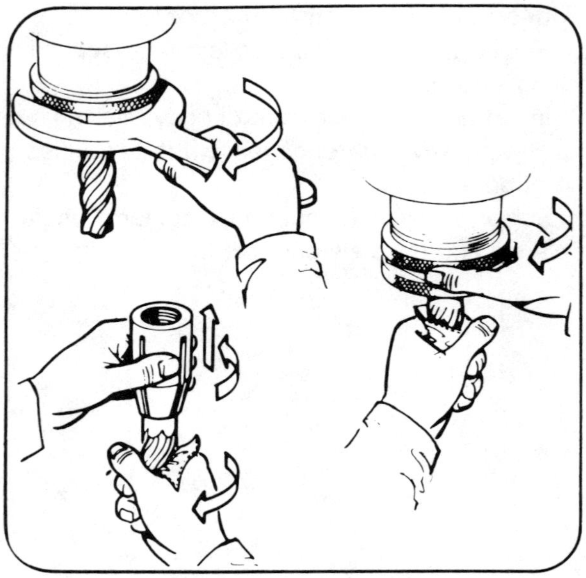

(b) Loosen locking nut with wrench and remove nut from arbor.
(c) Pull cutter out to remove sleeve and collet from arbor.
(d) Unscrew cutter and slide from collet.
(e) Clean all parts of chuck before storing away.

Note:
If changing cutter for one with same size shank:
(a) Loosen locking nut approximately half a turn.

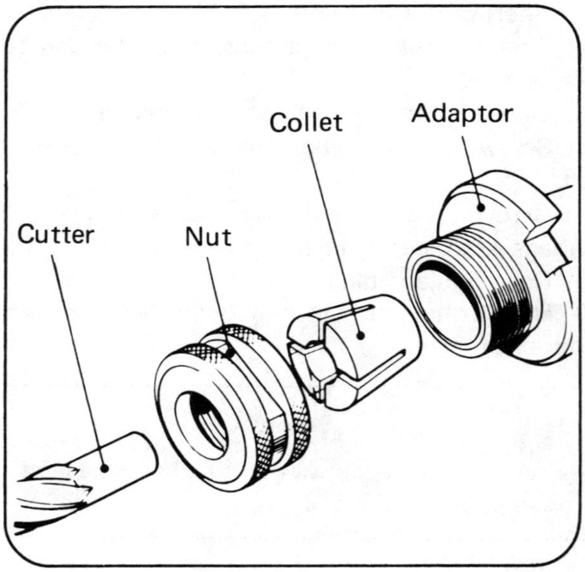

(b) Unscrew cutter and slide from collet.
(c) Insert new cutter in collet and screw up to center.
(d) Hand tighten locking nut.

Safety

Protect your hands by wrapping a rag around the cutter flutes.

Mounting cutters in collet chucks

1 Insert collet adapter in machine spindle

2 Insert collet in adapter
(a) Select collet same size as cutter shank.
(b) Clean bore of arbor and bore and outside of collet. Position collet in arbor.
(c) Screw nut on adapter until nut contacts face of collet.

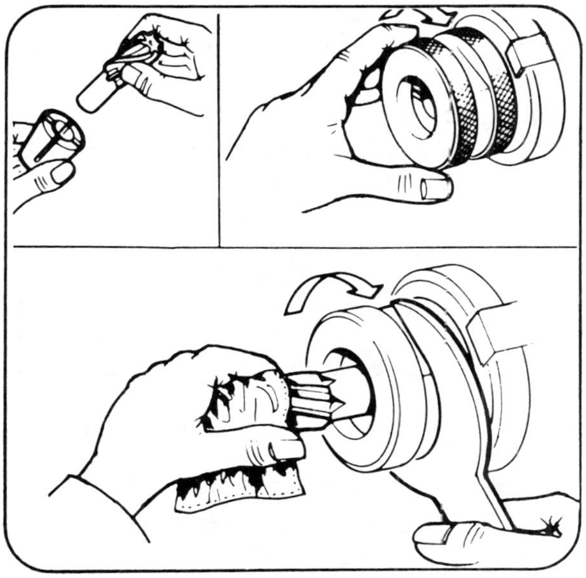

3 Mount cutter in adapter

(a) Insert cutter shank into collet.

(b) Slide shank into collet until end of shank contacts felt at rear of adapter.

(c) Hold cutter in position and tighten nut securely using correct wrench.

Arbor-mounted cutters: Vertical mill

Mounting a face mill on an arbor

1 Insert arbor in machine spindle

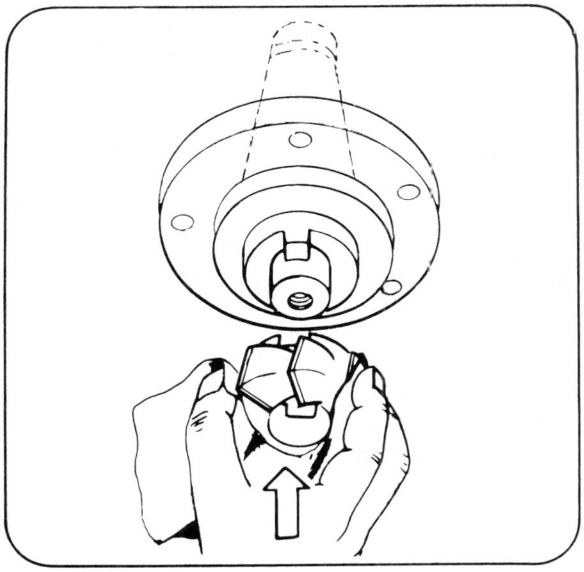

2 Mount cutter on arbor

(a) Unscrew locking screw from arbor.

(b) Clean mating parts of cutter and arbor.

(c) Slide cutter on arbor, rotating so that keys align with cutter slots.

(d) Feel for drive key engagement into cutter slot and check that faces contact properly.

> **Safety**
>
> Use a rag to protect your hands from the sharp edges of the cutter.

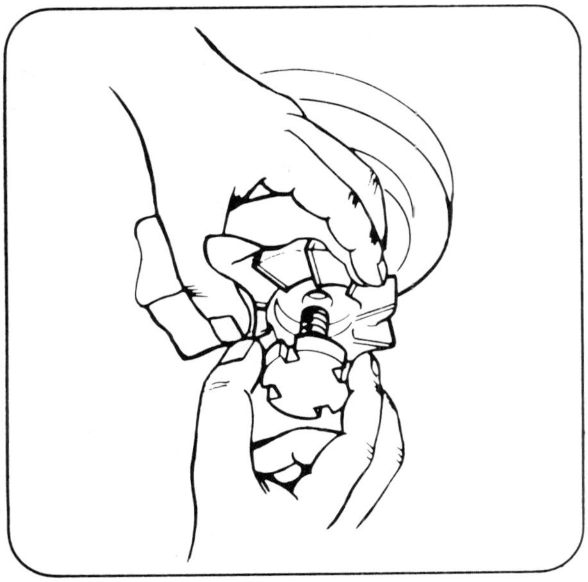

3 Tighten cutter on arbor

(a) Hold cutter in position on arbor and tighten locking screw finder tight.

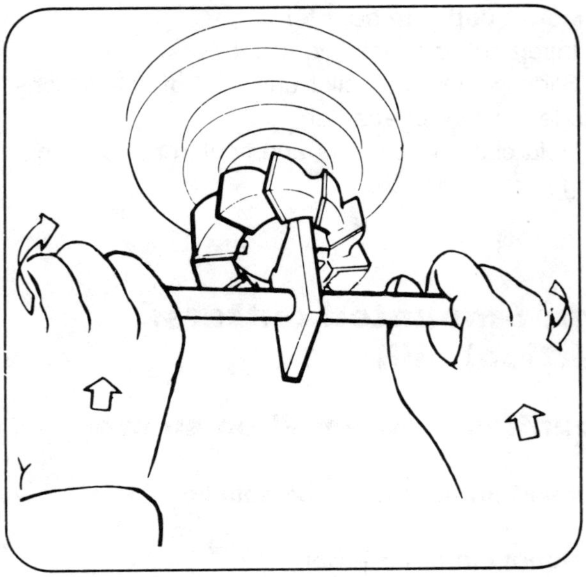

4 Finally, tighten locking screw

(a) Check that slots of key and slot of screw head are not worn.
(b) Engage slots of key in slots of screw head.
(c) Apply upward pressure to keep key engaged in slots and tighten screw.

Safety

Worn keys, or screws with worn slots, must not be used.

Shell end mills are mounted in a similar way except

1 A socket head screw is used as locking screw.

2 The correct hex key must be used.

3 A washer must be used between head of locking screw and shoulder of cutter.

Mounting cutters on a stub arbor

1 Insert arbor in machine spindle

2 Prepare arbor

(a) Hold bushings in position and remove securing nut.
(b) Remove collars and key from arbor.
(c) Clean collars and arbor faces.

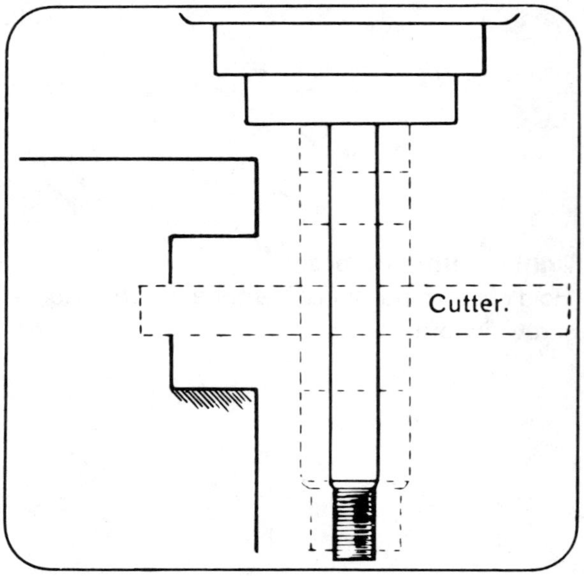

3 Decide cutter position on arbor

(a) Refer to drawings and workpiece to decide best position of cutter to obtain maximum possible rigidity while ensuring that arbor and workpiece will be clear of any obstruction.

Note:
The closer the cutter is positioned to the nose of the spindle, the greater the rigidity obtained.
(b) Decide number and width of collars to position cutter in correct position on arbor.
(c) Spacing collars should be lightly stoned before installing. This will aid in preventing cutter side runout.

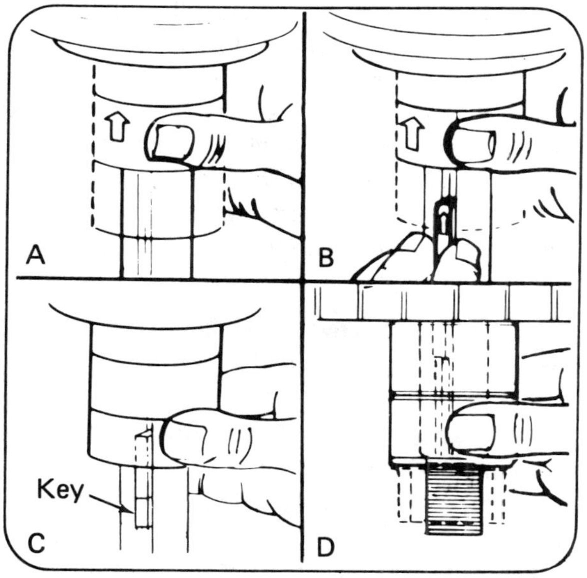

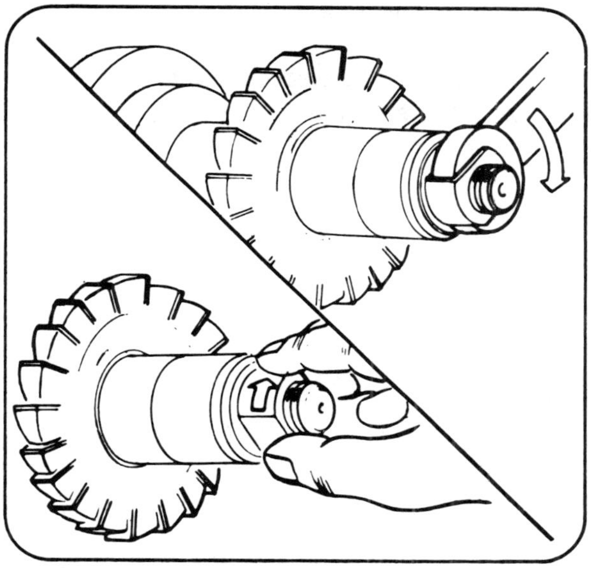

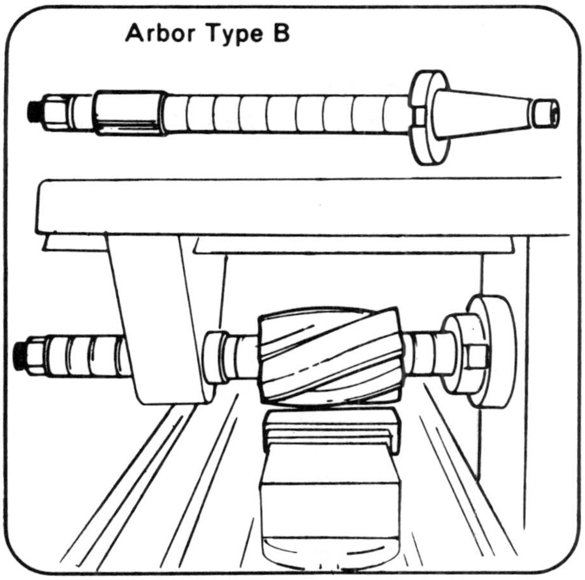

4 Mount cutter on arbor

(a) Slide collar or collars up to shoulder of arbor, one less than number required to position cutter (A).
(b) Fit key in arbor keyway, ensuring that key will extend into collars at each side of cutter (B).
(c) Align keyway in collar to key in arbor and slide collar on arbor (C).
(d) Align keyway of cutter to key and slide cutter up to collars, ensuring that cutting edge is facing in correct direction (D).

Note:
Thin cutters may be held between two collars for easier sliding onto the arbor.
(e) Align keyway and slide collar(s) onto arbor.
(f) Visually check that collar just extends onto threaded part of arbor, but leaves sufficient thread to engage nut fully.

5 Tighten cutter on arbor

(a) Hold bottom collar up to keep cutter in position.
(b) Engage nut on thread and finger tighten. Moderately tighten nut with wrench.

Arbor-mounted cutters: Horizontal mill

Mounting cutter on long arbor: arbor selection

Use arbor type B when the top surface of the workpiece will pass between the arbor flange and the support.

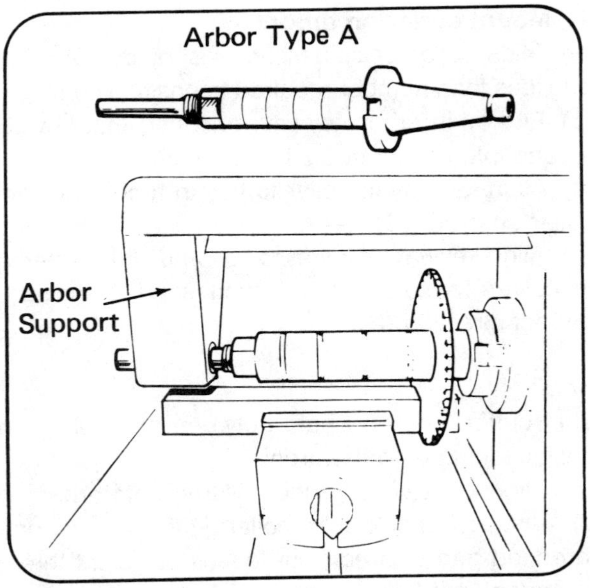

Type A arbors are used to enable the top face of a long workpiece to be raised to just below the diameter of the spacing collars without hitting the arbor support. If arbor B were used in this instance, exceptionally large diameter cutters would be required. The arbor nut may be either left- or right-hand thread. If possible, select an arbor on which the nut tightens in the opposite direction to the required cutter rotation.

Mounting a single cutter on a long arbor type B

1 Mount arbor in machine spindle

(a) Set machine in low gear.

(b) Clean taper bore of spindle and taper shank of arbor.

(c) Hold arbor approximately parallel to table surface.

(d) Insert arbor in machine spindle, rotating arbor to align drive keys to arbor slots.

2 Tighten arbor in machine spindle

(a) Hold arbor shank in position in spindle.

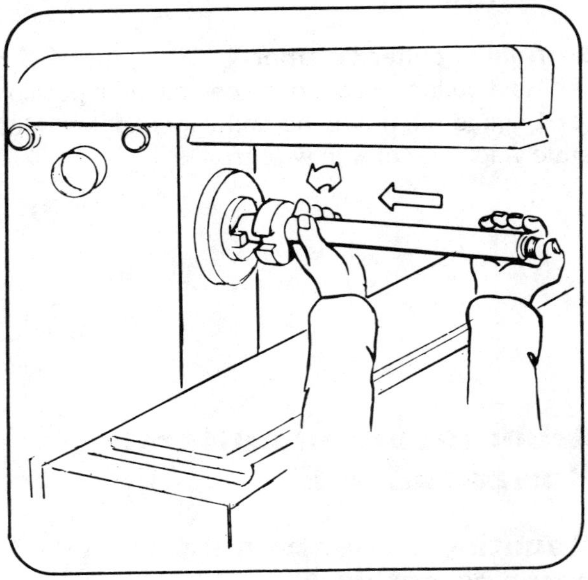

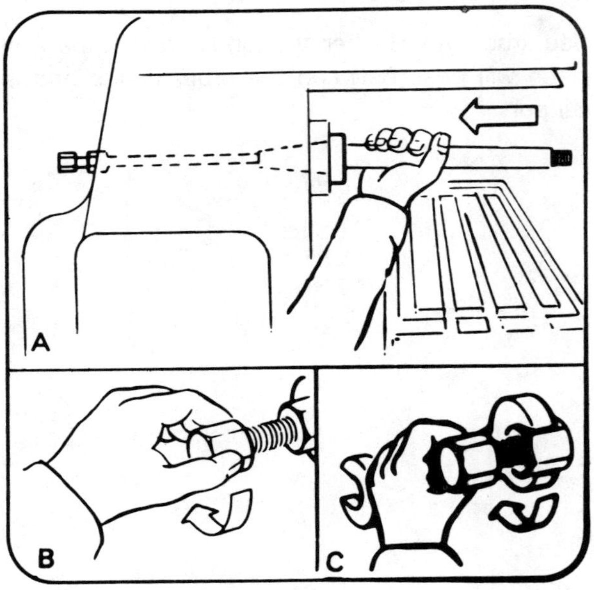

(b) Screw drawbolt well into arbor. Tighten drawbolt locking nut with wrench.

3 Decide cutter position

(a) Hand crank machine table in, to position workpiece close to rear of machine.

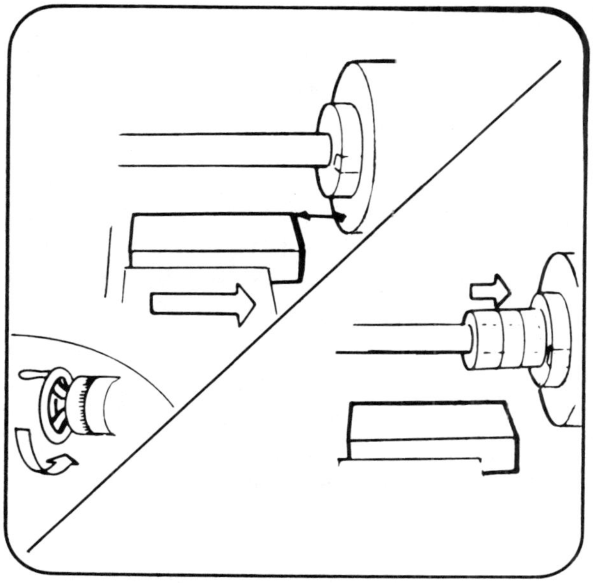

(b) Visually check that workpiece or work-holding device will clear front of arbor flange.
(c) Clean faces of collars and slide them onto arbor, pushing collars up to arbor flange. Stone collars to remove any burrs.
(d) Visually check buildup of collars to see that outer face is level with, or extending beyond, face to be machined.

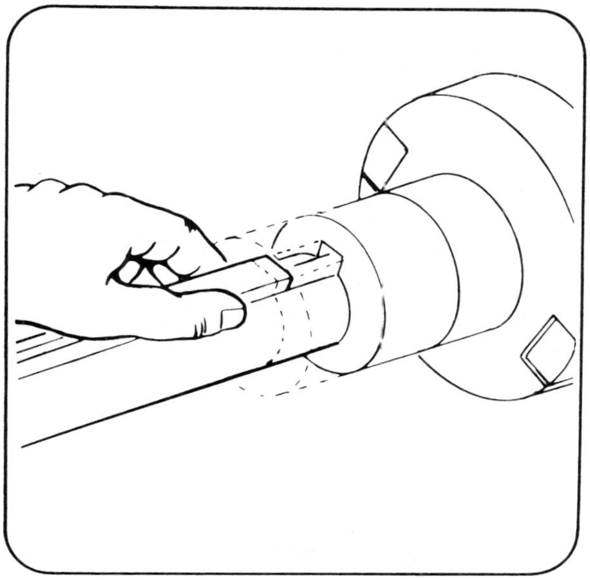

4 Fit key in arbor and mount cutter on arbor
(a) Remove outer collar from arbor.
(b) Fit key into arbor keyway, feeling for snug fit of key.
(c) Visually check that key will extend into collar on each side of cutter.
(d) Replace collar on arbor, rotating it to align keyway to key.
(e) Visually check that cutter is facing in direction of spindle rotation and slide cutter onto arbor.

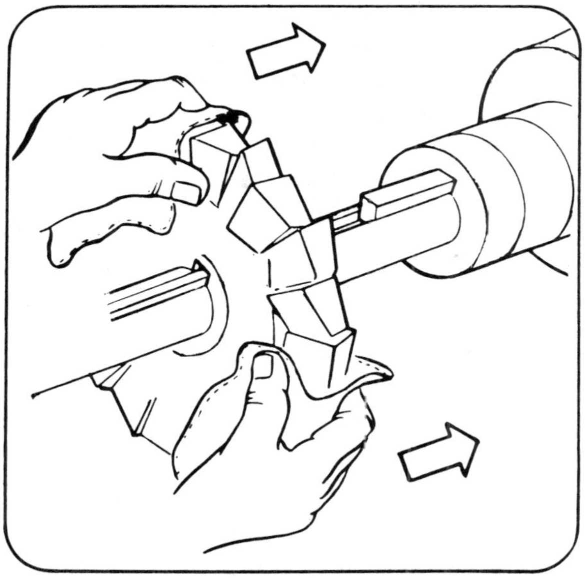

(f) Rotate cutter, to align keyway to key, and push up to collars.

Safety

Use rags to protect your hands from the sharp edges of the cutter.

Note:
When mounting thin saws on an arbor:
1 Omit fitting the key in the arbor. This allows the saw to slip if it gets trapped in the workpiece.
2 Trap the saw between two collars when sliding it onto the arbor. This stops the saw from tilting and makes it easier to slide onto the arbor.

5 Position arbor-bearing collar

(a) Check position of cutter to face to be machined.

(b) Adjust table setting if necessary to correct positioning.

Note:

If more than one part of the workpiece is to be machined, move the table out to position the cutter to the part of the workpiece to be machined that is closest to the spindle nose.

(c) Slide additional collars onto arbor, rotating as necessary to align keyway to key.

(d) Visually check buildup of collars, until outer face extends beyond front end of workpiece.

(e) Slide support-bearing collar up to collars on arbor.

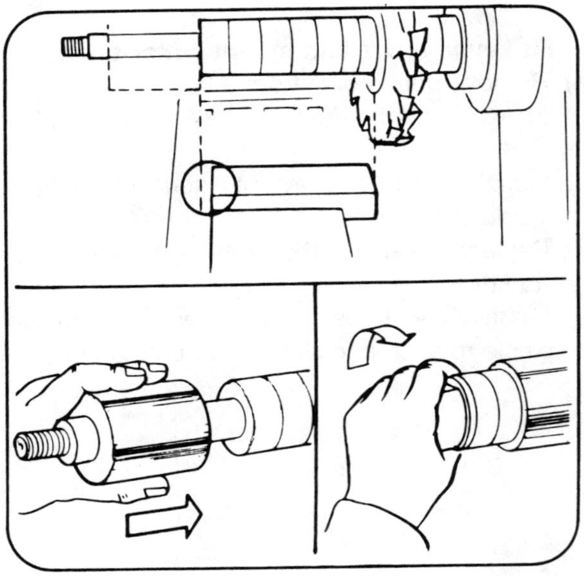

Note:

The closer the support-bearing collar is positioned to the cutter, the greater the rigidity obtained when the support is positioned.

(f) Slide additional collars onto arbor until they just extend over threaded part of arbor.

Note:

One or more of these collars should be keyed to the arbor if the cutter has not been keyed. This ensures that the nut will not tighten if the cutter slips.

(g) Visually check that sufficient thread extends for nut to be fully engaged. Engage nut and finger tighten.

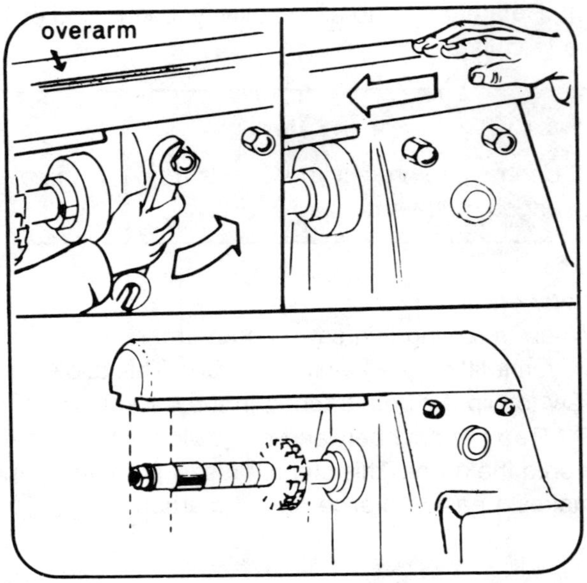

6 Engaging arbor support

(a) Loosen overarm securing nuts.

(b) Slide overarm out while visually checking that end extends beyond outer end of bearing bushing.

(c) Clean mating parts of overarm and support.

(d) Visually align mating parts of support and overarm.

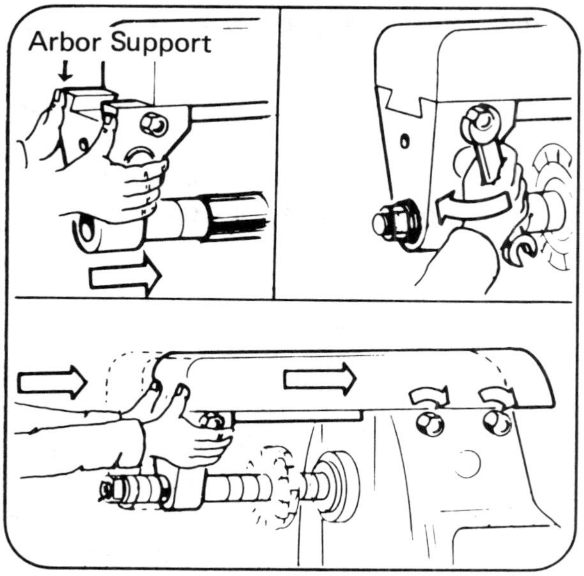

(e) Slide support on overarm to a position at which front support is even with the end of the overarm.
(f) Tighten nut to secure support to overarm.
(g) Slide overarm in so that support bearing is centered on arbor journal.
(h) Tighten overarm securing nuts tightly using appropriate wrench.

Note:
Never tighten the arbor nut unless the arbor support is in place and supporting the arbor. Attempting to tighten the nut without the arbor support in place may bend the arbor.

7 Final tighten the cutter on arbor
Position the wrench on the arbor nut with the handle approximately vertical and tighten the nut securely.

Safety

Do not use a worn wrench on the arbor nut.

8 Check true running of cutter
(a) Check that cutter is clear of workpiece.
(b) Start machine spindle and run cutter at moderate speed. Be sure that support bushing fit to arbor-bearing collar is properly adjusted and lubricated.

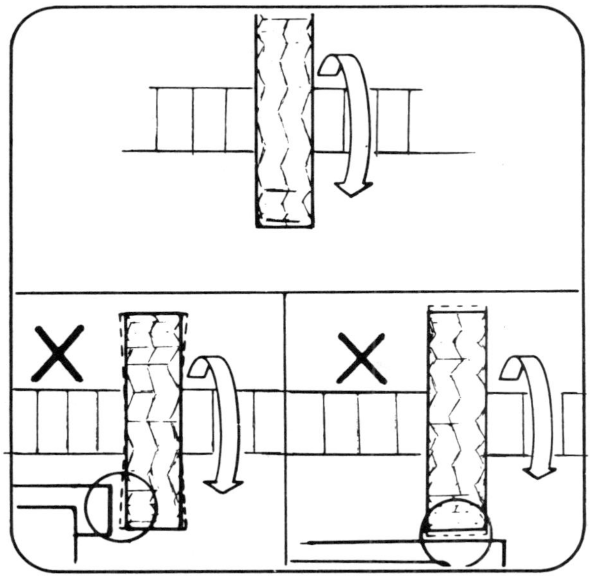

(c) Visually compare side of cutter to a fixed part of machine or workpiece to see if cutter wobbles.
(d) Visually compare periphery of cutter to a fixed part of machine or workpiece to check concentricity to spindle rotation.

Note:
If the cutter wobbles:
1 Check that the faces of the spacing collars, cutter, arbor, and securing nut are clean and undamaged.
2 Check that the arbor is correctly mounted and the tapers are clean and undamaged.
3 Check that the faces of the spacing collars are parallel.
If the wobble is still present:
1 The arbor may be bent. Check concentricity of arbor.
2 The cutter may be incorrectly ground.

Mounting cutters on a long arbor, type A

Cutters are engaged on type A arbors in exactly the same way as on arbor type B except that no bearing bushing is required. The pilot diameter runs in a bearing that is built into the arbor support. In addition, you must:

1 Visually check that the nut does not extend onto the pilot diameter of the arbor.

2 Visually check that the bearing in the support is clear of the arbor thread.

3 Visually check that the bearing in the support does not extend beyond end of pilot diameter.

4 Key one or more of the collars between the cutter and the securing nut, if the cutter is not keyed to the arbor.

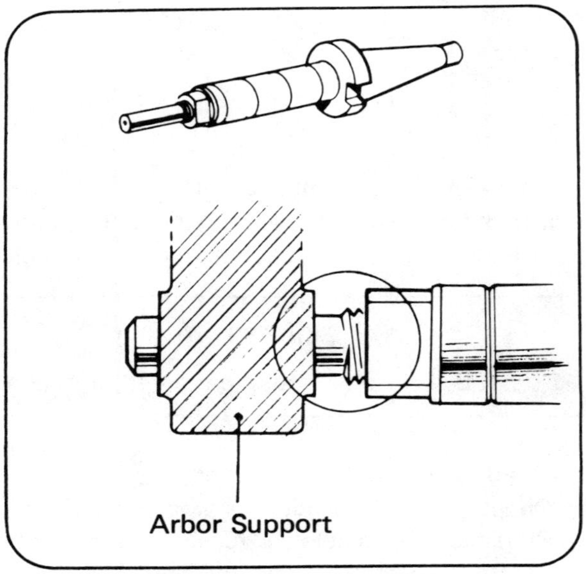

Setting single-point cutters

The cutting edge of a single-point cutter should always be set on the centerline of the tool holding device. The smaller the diameter machined by the tool, the more important the accuracy of this setting becomes.

1 Set cutter holding device in machine spindle

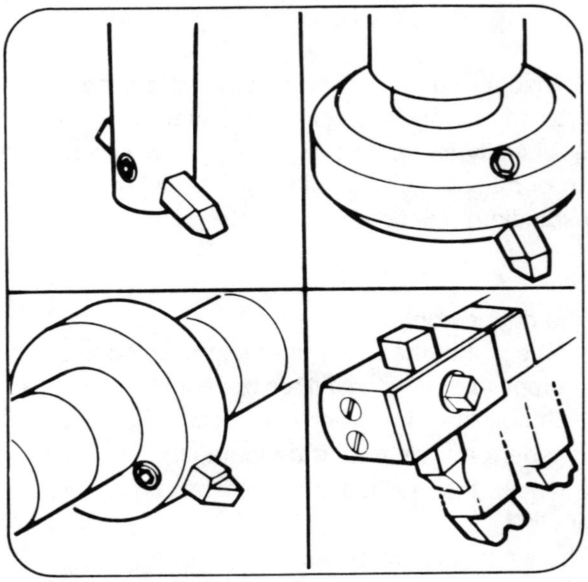

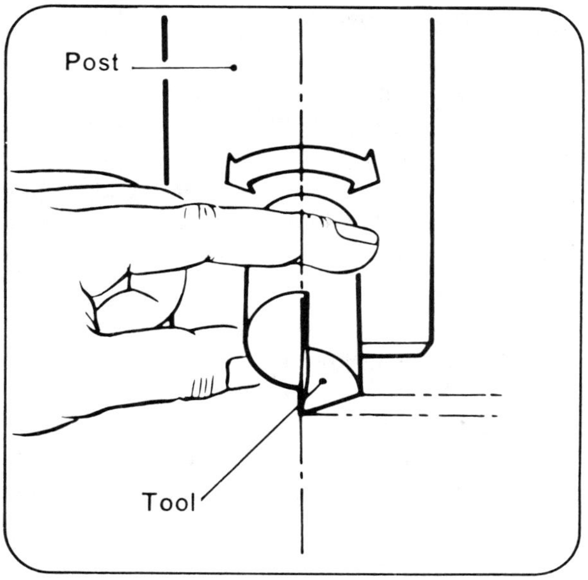

2 Position cutting tool in holding device

(a) Rotate spindle to position tool hole in holding device to front of machine.

(b) Mount tool in holding device with minimum amount extending.

(c) Visually check bottom of tool and rotate it to position front cutting edge lower than rear of tool.

(d) Visually check that front cutting edge is approximately on centerline of arbor.

(e) Hold tool in position and tighten securing screw with appropriate key, ensuring that tool does not move.

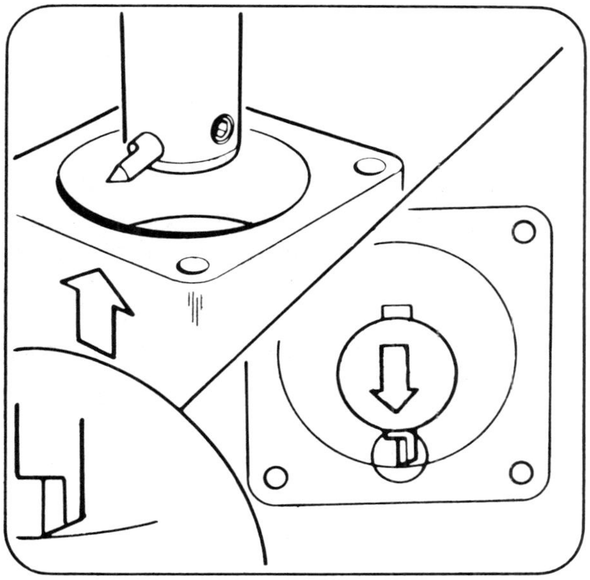

3 Check tool clearances

(a) Lightly touch cutting tool to workpiece.

(b) Visually check for back and side clearance between cutting tool and workpiece.

Mounting cutter gangs

The term **gang milling** is used if more than one cutter is mounted on the arbor. This permits machining several features at the same time.

Cutters used in gang milling should be paired and with opposing helices, so that the cutter thrust is neutralized. This avoids excessive wear and helps to eliminate chatter. The largest possible diameter arbor and the smallest possible diameter cutters should be used.

Cutter assembly

1 Position workpiece

(a) Set workpiece as near to rear of table as possible without overhang.

(b) Align workpiece in required plane, and clamp it.

(c) Hand move table in, as near rear of saddle as possible, leaving small amount of saddle movement for final adjustment.

2 Mount arbor in spindle

3 Fit key in arbor

(a) Rotate spindle to bring arbor keyway to the top.

(b) Visually judge position of cutters by relating workpiece to arbor.

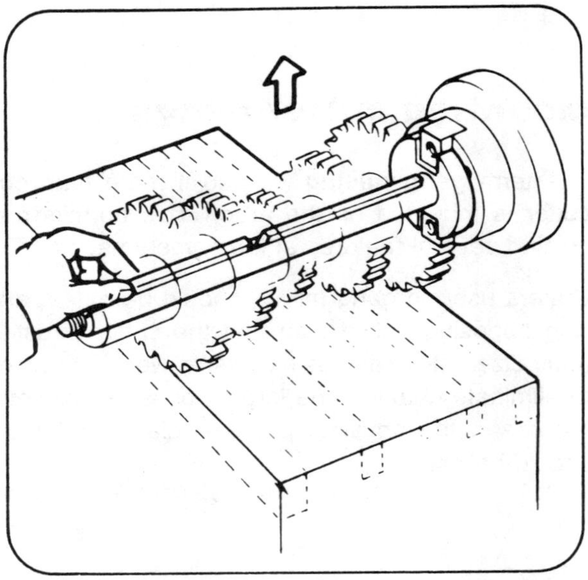

(c) Fit key in arbor to extend beyond each end of cutter positions.

4 Mount first cutter

(a) Select required collars to position first cutter to appropriate part of workpiece, but as near spindle nose as possible.

(b) Ensure that collar faces and arbor are clean. Engage collars on arbor, rotating them to align keyway to key, and push them up to arbor flange.

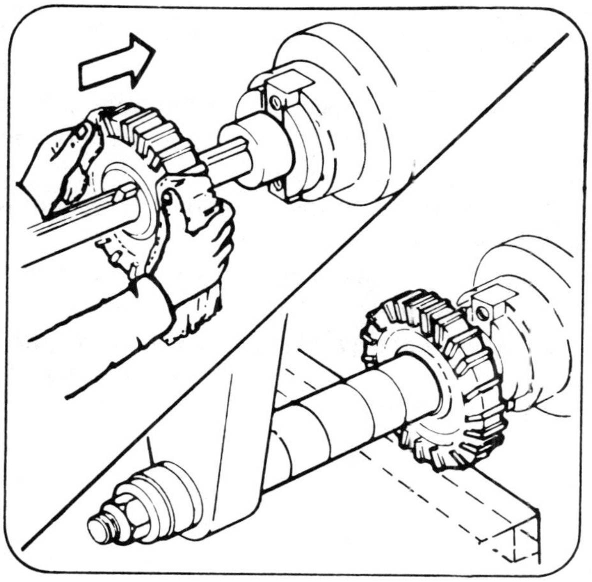

(c) Mount cutter on arbor, ensuring that direction of rotation is correct, and push up to collars.
(d) Build up collars to end of arbor, positioning support-bearing collar clear of workpiece. Finger tighten nut.
(e) Position overarm and arbor support on arbor-bearing collar and tighten cutter on arbor.
(f) Rotate cutter and check for true running.

5 Accurately set workpiece to cutter
(a) Start machine spindle and touch cutter to face of workpiece.
(b) Set light cut and traverse approximately ½ in. (12 mm) into workpiece.

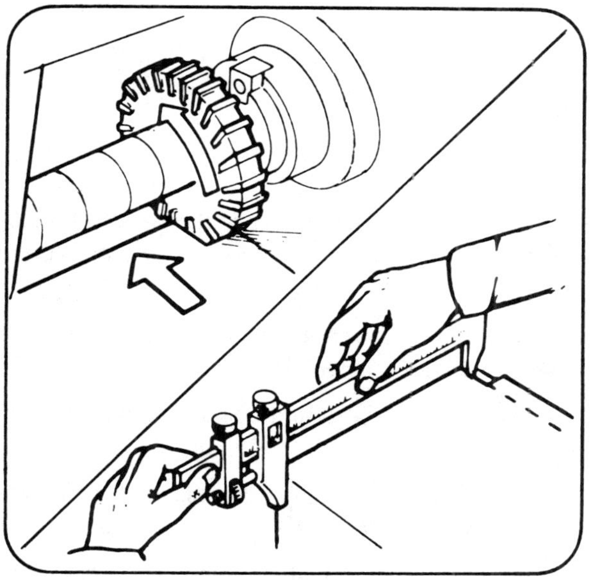

(c) Measure workpiece to check position of face.
(d) Adjust table setting to correct any error found.

6 Set second cutter
(a) Loosen arbor nut, disengage arbor support, and remove collars from arbor up to cutter.
(b) Build up required distance between cutters with spacing collars and shims.
(c) Position second cutter on arbor, ensuring that direction of rotation is correct.

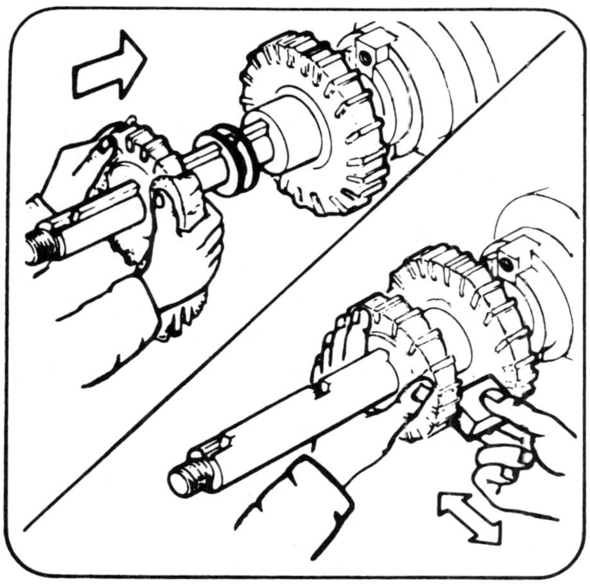

(d) Hold cutter hard up to collars and check distances between cutter teeth with gage blocks or other appropriate measuring tools.

Note:
Allow distance of approximately .002 in. (.05 mm) larger than finished dimension of workpiece because this amount will close in when cutters are properly tightened.
(e) Add or remove shims to adjust setting.

7 Check cutter setting accuracy
(a) Build up collars on arbor, position arbor support, and tighten cutters on arbor.
(b) Rotate arbor and check for runout of cutters.
(c) Start machine spindle and traverse approximately ½ in. (12 mm) into workpiece.

(d) Measure workpiece to check dimension and adjust distance between cutters if necessary.

8 Use of the intermediate arbor support

(a) Use intermediate support to keep unsupported length of arbor to a minimum.
(b) Select most convenient position for support to obtain maximum support.
(c) Visually check that support will not hit on workpiece or workholding device.
(d) Position arbor support in center of bearing bushing and securely tighten in position after final adjustment of overarm.

Review questions

1 Describe and demonstrate the procedure for mounting cutters directly to the machine spindle.

2 What safety precautions should be observed when handling milling cutters in general?

3 Describe and demonstrate how cutters are mounted in collets.

4 Describe and demonstrate how straight and taper shank cutters are mounted in the machine spindle.

5 Describe and demonstrate the mounting of cutters on arbors for vertical and horizontal milling.

6 What primary safety precaution must be observed when tightening an arbor nut on a horizontal mill?

7 What may happen if the correct procedure for question 6 is not followed?

8 How should arbors and collars be stored when not in use?

9 How should the cutting edge of single-point cutters be positioned relative to the centerline?

10 Describe and demonstrate gang cutter mounting.

UNIT SIX

Workholding Methods, Tooling, and Setups

Useful tips for workholding and setting

Workholding

1 Select bolts as short as possible, but with sufficient thread above the clamp and the washer to allow the nut to fully engage on the thread.

2 Position bolts as close as possible to the workpiece.

3 Position the clamp so that the length from the bolt to the workpiece is less than the length to the clamp support.

4 Support the rear of the clamp until the clamp is level or slightly higher at the supported end.

5 Ensure that the clamps and supports will clear the cutters and arbors.

6 Always clamp on a solid part of the workpiece. Never clamp over unsupported parts of the workpiece.

7 Protect finished surfaces against damage by inserting soft material between the clamp and the workpiece.

8 Always clamp the workpiece in at least two places.

9 Before moving a clamp after partial machining, ensure that one or more other clamps are holding the workpiece in position.

10 Position clamps as near the ends of the workpiece as possible.

11 When clamping through a bore, set stops to prevent the workpiece from rotating.

12 Lightly tighten all clamps before final tightening, using even pressure on each clamp alternately.

13 When clamping or securing workpieces, such as thin plate, or materials that distort easily, clamps or a vise should always be released and carefully retightened to take up the spring before the final light finishing cut is taken.

14 When mounting long workpieces between a dividing head and a tailstock, support the workpiece midway between the ends, with an adjustable Vee jack.

Setting

1 Cleanliness is essential. Always clean both machine and tools before setting up.

2 Check for and remove burrs on all tools and workholding equipment.

3 Set workholding equipment to position the workpiece midway between the ends of the table, if possible.

4 Avoid excessive overhang of the workpiece or workholding equipment.

5 Set indexes of all indexing equipment at zero or 90° before setting and before removing from the machine after use.

6 When setting a workpiece or workholding equipment in more than one plane, roughly set in each plane before the final setting.

7 Set the workpiece or workholding device to allow the table to be positioned as near as possible to the vertical column.

8 Position the workholding device and/or workpiece to ensure that the cutter thrust is toward a fixed location, or set stops.

9 Whenever possible, ensure that the workpiece is so positioned that chips removed are directed away from the operator.

10 Align dividing head and tailstock before mounting the workpiece.

11 Check the workpiece for accurate setting before machining.

12 Ensure that index sleeves are tight and mark the direction of free movement (backlash) before indexing the machine table.

13 Recheck all calculations and table settings before starting to machine the workpiece.

14 A simple setup is usually the best one.

Slinging and lifting a vise

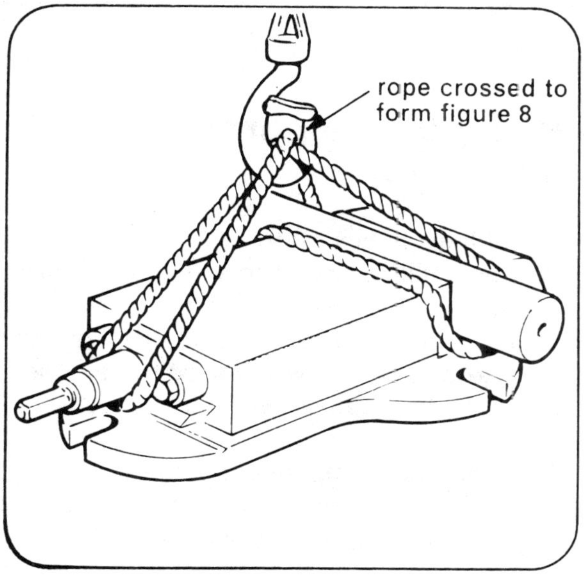

1 Position and clamp length of bar in vise

(a) Extend bar 3 to 4 in. (75 to 100 mm) beyond both ends of vise jaws (ensure that bar is thick enough to carry weight of vise).

(b) Tighten vise jaws hard on to bar.

2 Position rope sling

(a) Check that safe working load of rope sling is adequate for weight of vise.

(b) Inspect sling for signs of wear or damage.

(c) Cross sling to form figure 8.

(d) Position one loop under each end of bar.

(e) Position other loop under handle end of vise screw.

(f) Position crane hook through and under crossed sling, approximately centered on vise body.

Safety

A rope crossed on top of the hook prevents the rope from slipping.

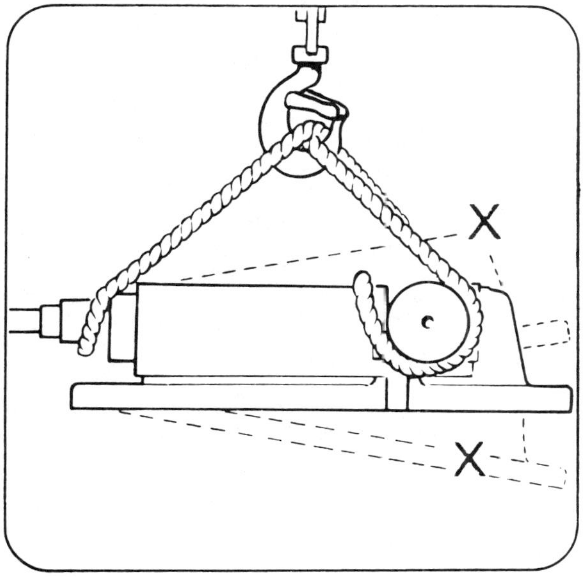

3 Lift vise

(a) Ensure that crane is directly over vise.

(b) Raise vise just off floor or table.

(c) Vise should lift level. If either end tilts up, lower vise and adjust hook position toward lower end. Adjust until vise lifts level.

(d) Raise vise to just above machine table level.

Safety

Stand clear when lifting with a crane.

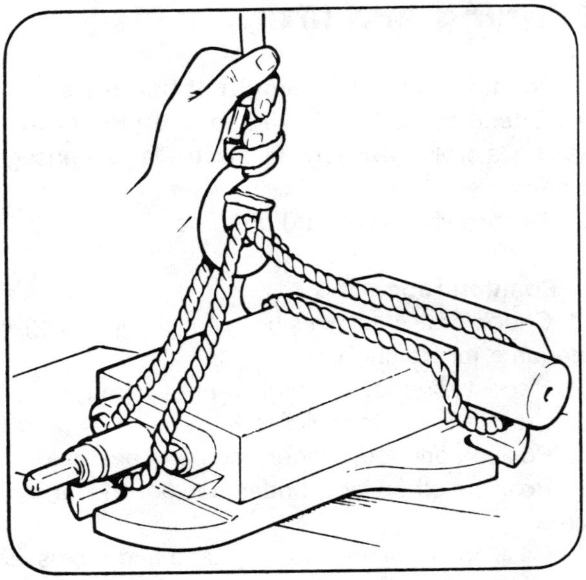

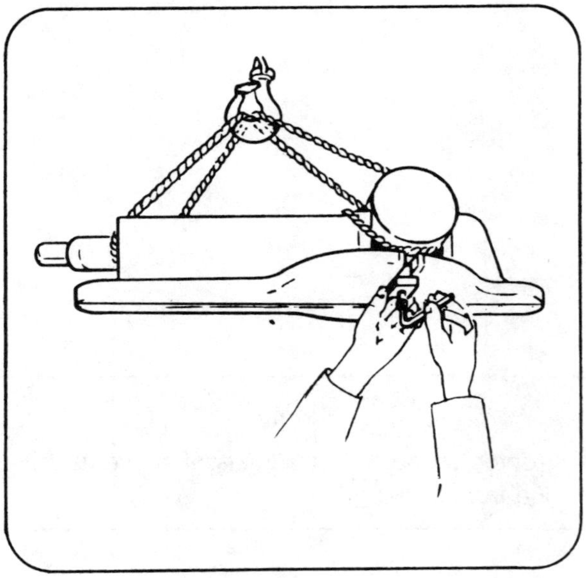

4 Position vise on machine table

(a) Wipe base of vise clean. Stone off any burrs on vise or machine table.
(b) Position vise over table.
(c) Lower vise gently onto table.
(d) Ensure that vise is securely seated on table before removing hook and sling.

Safety

The crane hook should be pushed into a safe position clear of aisles and working area.

Setting a vise parallel using keys

1 Insert keys in key slots

(a) Check that keys are clean and free from burrs.
(b) Clean key slots.
(c) Insert keys in slots, and ensure that cap screw holes in keys are in line with tapped holes in vise.
(d) Insert socket head screws and tighten.

2 Position and clamp vise to table

(a) Lower vise to machine table with keys approximately centered over table slots.

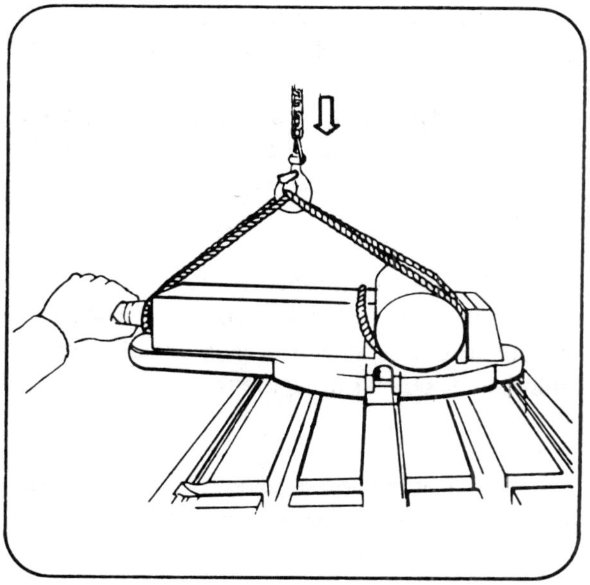

(b) Engage keys in table slots.
(c) Check vise to ensure a good seating on table, without movement.
(d) Tighten securing bolts.

Note:
If the vise can be lifted by hand, tilt it onto its side and insert the keys before lifting to the machine table.

Setting a vise parallel without keys

1 Position vise approximately parallel
(a) Check that vise is clean and free of burrs.

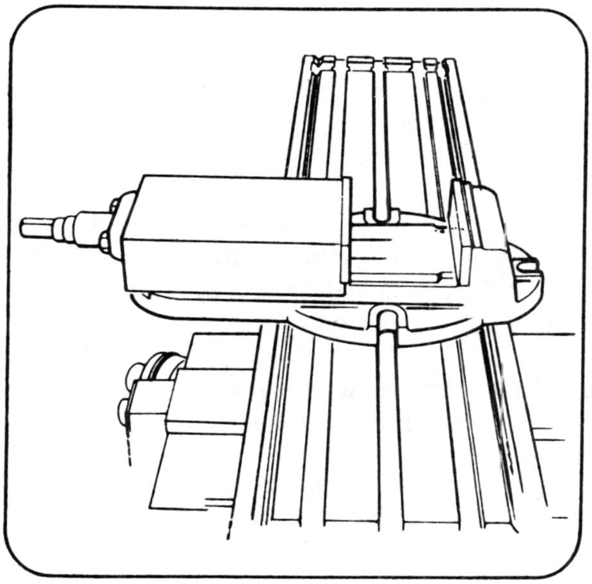

(b) Position vise to obtain maximum support from machine table.
(c) Align vise slots approximately centered over table slot.

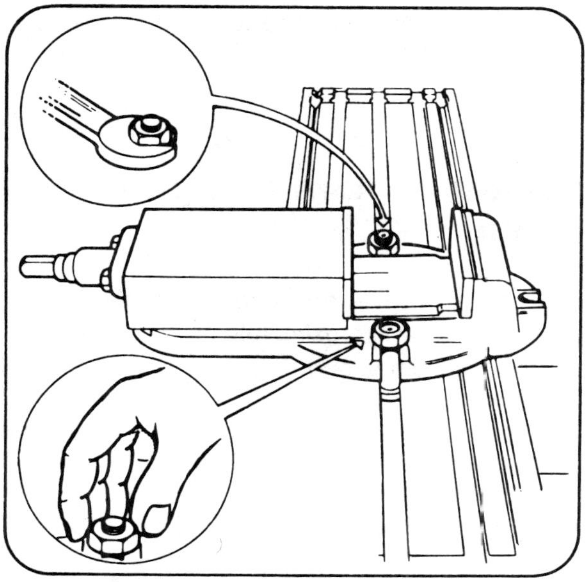

2 Lightly clamp vise to table
(a) Ensure clearance between bolts and vise slots to allow for movement.
(b) Form pivot point by gently tightening one securing bolt.

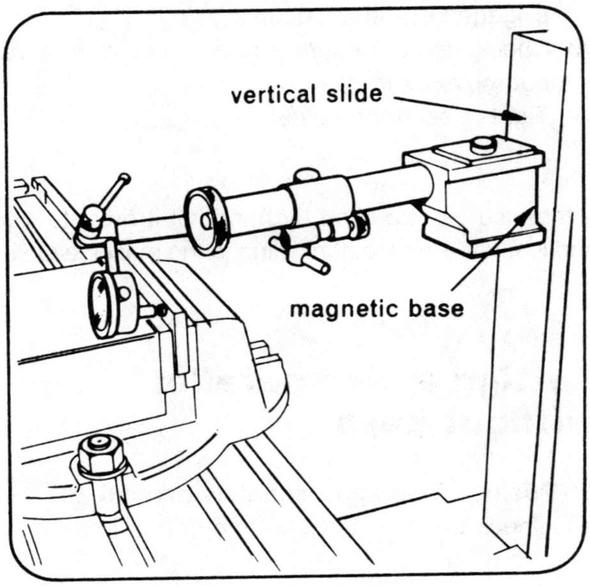

3 Position dial indicator

(a) Ensure that magnetic base is firmly attached to vertical slide.
(b) Engage dial indicator stylus on pivot point end of fixed vise jaw or onto a parallel held in vise.
(c) Keep dial indicator body clear of vise, and stand clear of it.

4 Check vise alignment

(a) Move/traverse table longitudinally to pass dial indicator to opposite end of jaw, visually checking dial indicator pointer movement.

Note:
Use the dial indicator readings near the extreme ends of the jaw to obtain greatest accuracy.

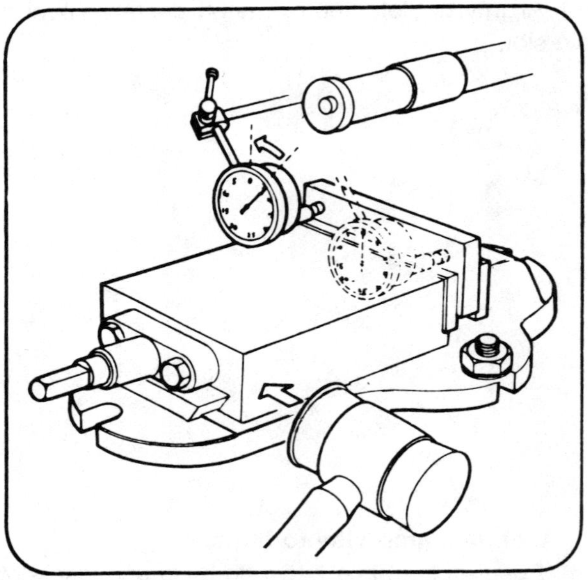

(b) Tap side of vise body at handle end of vise to correct any error.

Note:
NEVER tap the vise toward the dial indicator unless the stylus has been removed from the jaw. Tapping when the stylus is in the jaw may damage the delicate indicator mechanism.

5 Final clamping of the vise

(a) When clamping, be careful to avoid movement of vise.
(b) Check alignment after clamping.
(c) Reset vise if any movement has occurred.

Note:
The same method is used to set the vise square across the table.

Setting a plain vise at an angle

1 Position vise at approximate angle

(a) Center vise approximately on machine table to obtain maximum support.

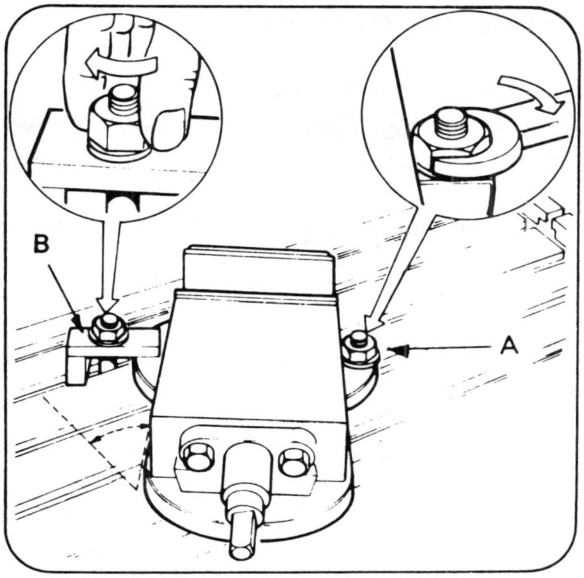

(b) Position a table bolt (A) in vise slot before setting vise at an angle.

(c) Set vise visually, to approximate angle required.

(d) Tighten table bolt (A) to form pivot point.

(e) Position a clamp (B) on vise flange on opposite side to table bolt and finger tighten.

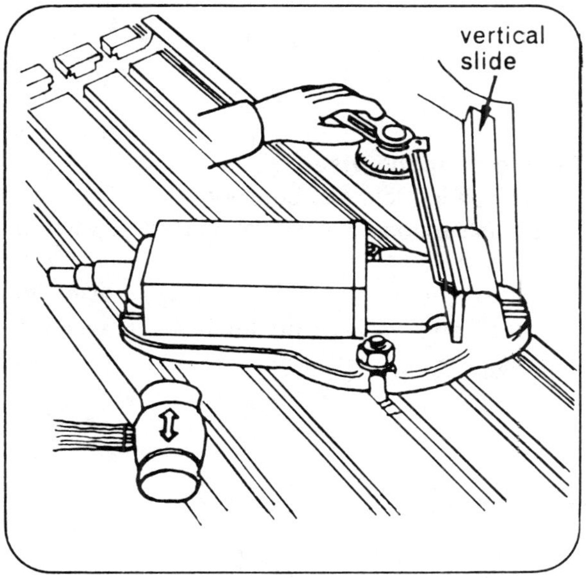

2 Setting the angle

(a) Set protractor to required angle.

(b) Clean base of protractor and vertical slide of machine.

(c) Hold protractor base firmly against vertical slide, parallel to table, and slide it along until its blade makes contact with vise jaw.

(d) Tap vise until full contact is made between protractor blade and vise jaw.

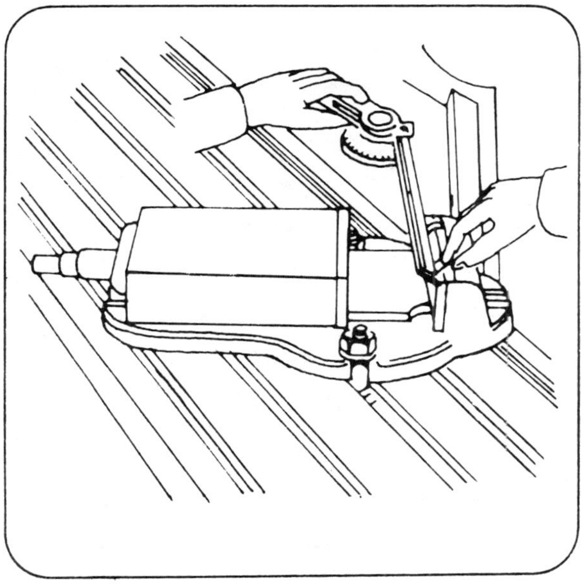

3 Accurately set vise

(a) Position pieces of paper on extreme ends of face of vise jaw.

(b) Slide protractor to make light contact with papers and pull each paper lightly to feel for movement.

(c) Tap side of vise where paper is loose and adjust until both papers are trapped.

4 Final clamp vise

(a) When clamping be careful to avoid movement of vise.

(b) Check angular setting after clamping.

Setting vises in line

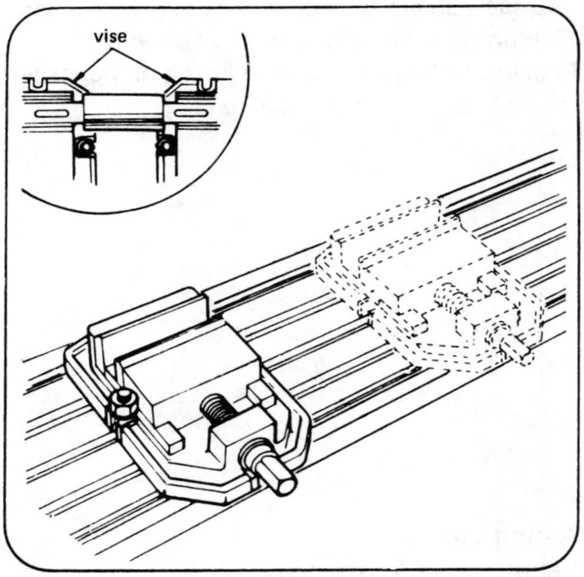

1 Position and set first vise parallel

(a) Refer to drawing to select best position for vises to enable full machining operation(s) to be carried out.

(b) Place one vise in required position and set it parallel.

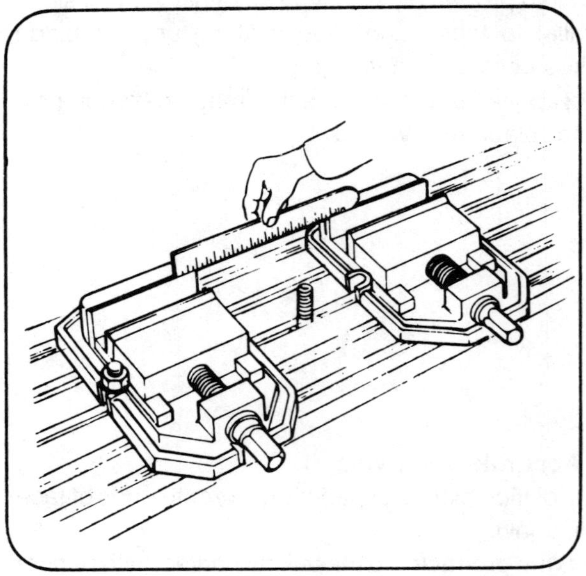

2 Position second vise

(a) Place bolt in table slot between vises.

(b) Position second vise, visually aligning jaws to first vise.

(c) Judge or measure distance between vises to ensure that jaws will not obstruct any machining operations.

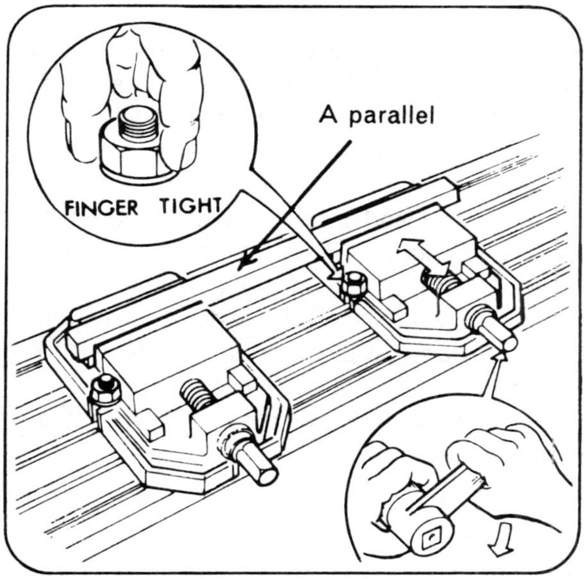

3 Align vises

(a) Place long sturdy parallel between jaws of both vises to engage at least two-thirds the length of both vise jaws.
(b) Tighten jaws of previously set vise to grip parallel securely.
(c) Pull second vise manually until fixed jaw lightly touches parallel. Tighten jaws securely.
(d) Tighten table bolts to secure vise.

Note:
Where high accuracy of alignment and parallelism to table movement is required, mount a dial indicator and check over the full length of the parallel. Adjust vise positions as necessary to achieve a zero indicated runout.

Setting a swivel vise at an angle

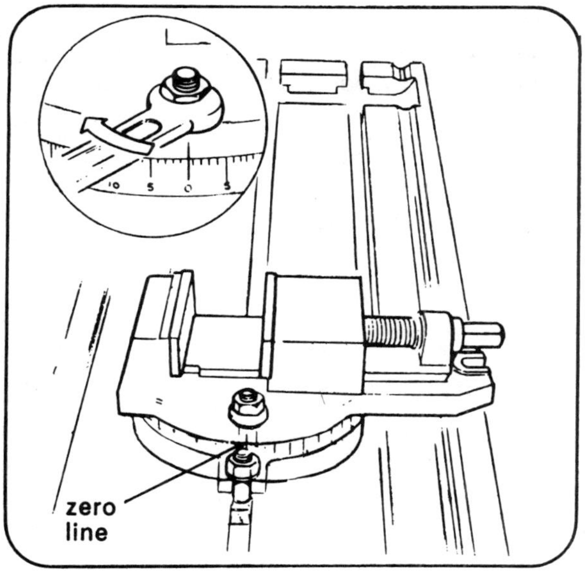

1 Prepare vise for setting

(a) Align setting line of vise to zero line on scale.
(b) Tighten bolts to clamp swivel part of vise to swivel base.

2 Clamp vise and set square or parallel

(a) Refer to part and/or drawing for best vise position.
(b) Set vise parallel to machine table.

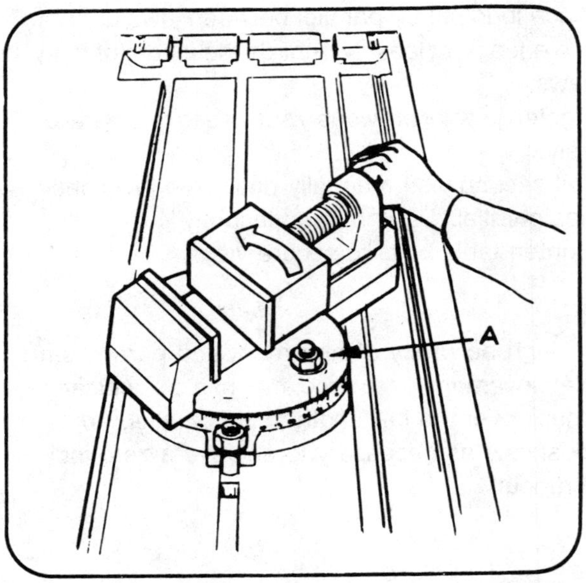

3 Set vise at required angle

(a) Unclamp swivel part of vise only (nut A).

(b) Swivel vise until setting line on vise is aligned with required graduation on swivel base scale.

Note:

A vise set parallel and swiveled to 30° angle will produce a 30° angle on the part if a cut or cuts are taken with table movement, and 60° with table movement if a cut or cuts are taken with saddle movement. This will be reversed if the vise is set square and swiveled to 30°.

(c) Clamp vise securely to swivel base (nut A), being careful not to move vise.

Setting a universal vise at an angle

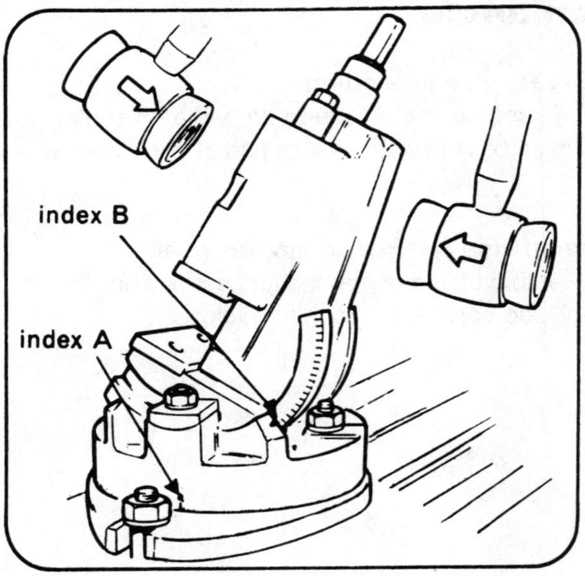

1 Set vise square or parallel

(a) Ensure that both index readings are set at zero (A and B).

2 Tilt vise at angle

(a) Loosen vertical swivel retaining bolts.

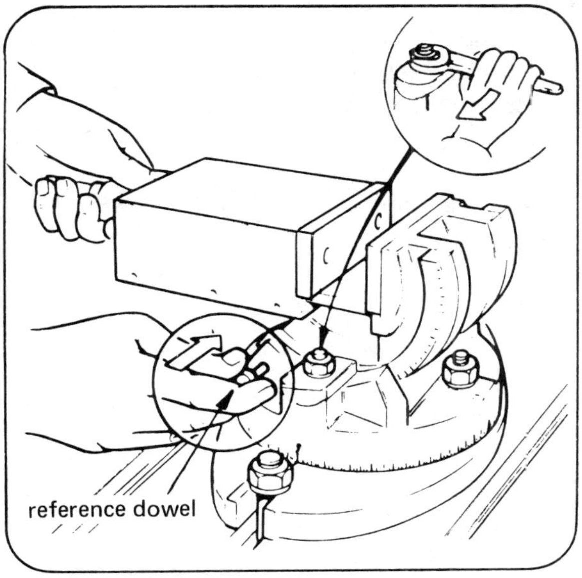

(b) Pull reference dowel out to clear zeroing hole.
(c) Tilt the vise up until setting line is in line with required graduation on angular scale.
(d) Tighten swivel retaining bolts lightly.
(e) Tap vise up or down with soft hammer until setting line is obtained.
(f) Tighten retaining bolts.

Safety

Be careful of hands when loosening and tilting a vise.

Note:
For a more accurate setting, the angle may be checked with a protractor or sine bar. The vise may be swiveled on its base, thus obtaining compound angles. All angular indexes should be reset at zero before removing the vise from the machine table.

Using soft jaws

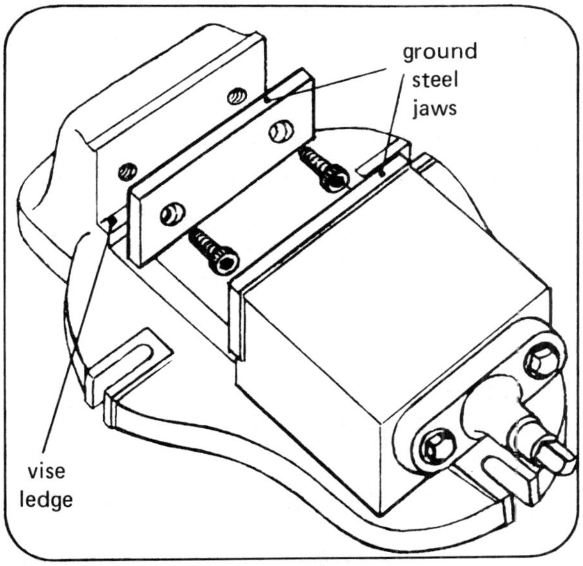

Machine vises, with the sliding jaw operated in parallel with the fixed jaw, are usually fitted with hardened ground steel jaws. These are held in position by screws with heads sunk below the level of the contact faces.

Soft jaws are interchanged with the hardened steel jaws, held by the same screws, and form an integral part or extension of the vise. They may be made of both metallic and nonmetallic materials capable of being machined and rigid enough to support the workpiece.

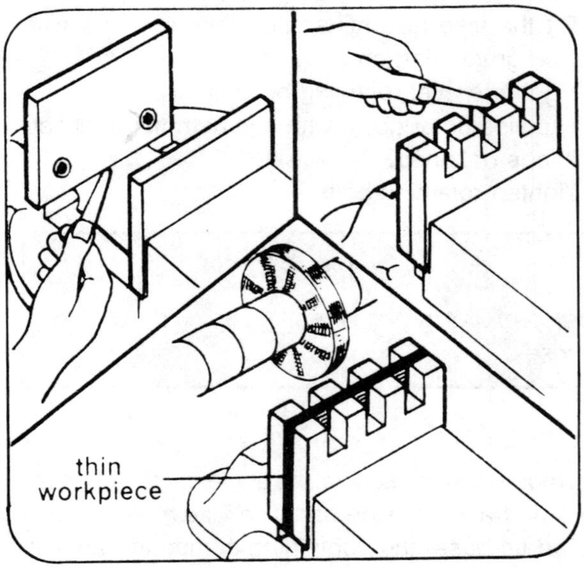

1 Supporting thin workpieces

(a) Fit extended jaws, preferably of same type material, and check with feeler gage that base edge is in contact with vise ledge.
(b) Close vise jaws, set up cutter(s), and mill as required.
(c) Clean cutters and clean and deburr jaws, insert workpieces, and mill.

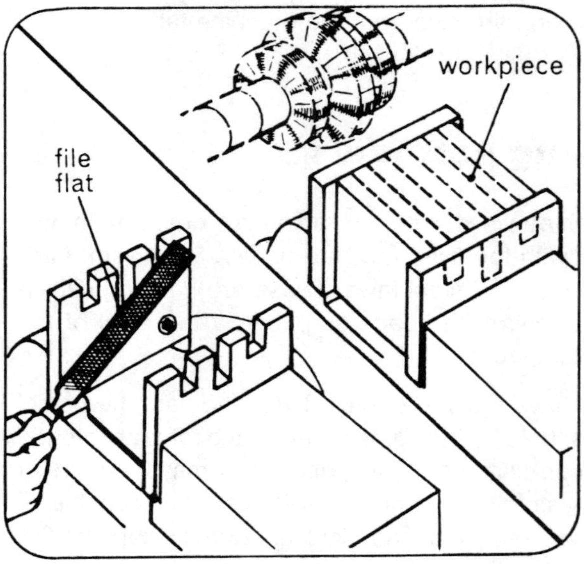

2 Protection of hard plastics

(a) Select material for soft jaws that will not mark finish of workpiece.
(b) Set up and hold workpiece in vise.
(c) Complete milling operation.
(d) Remove workpiece.
(e) File inside faces of jaws to remove any protruding burrs.

Note:
Avoid chamfering any corners. Square corners are essential to give full support to a brittle workpiece.

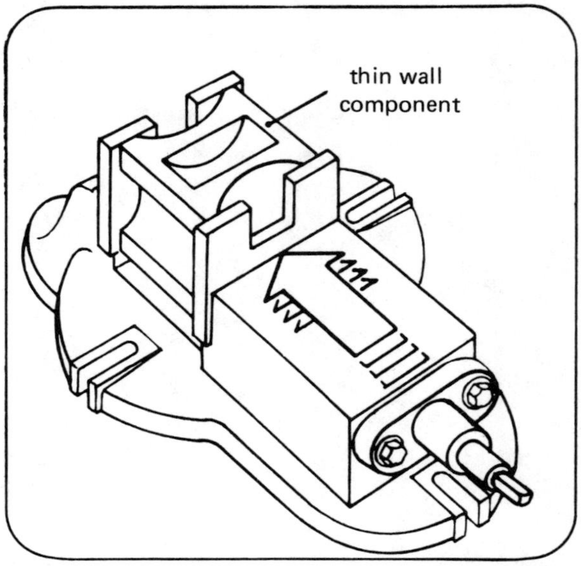

3 Protection for ductile plastics

(a) Soft jaws may be made of metal or a nonmetallic rigid material.
(b) Fit jaws, close vise, set up cutters, and mill as required.
(c) Clean cutters and clean and deburr jaws, insert workpiece, and tighten vise just enough to hold, taking care to prevent compression or distortion.

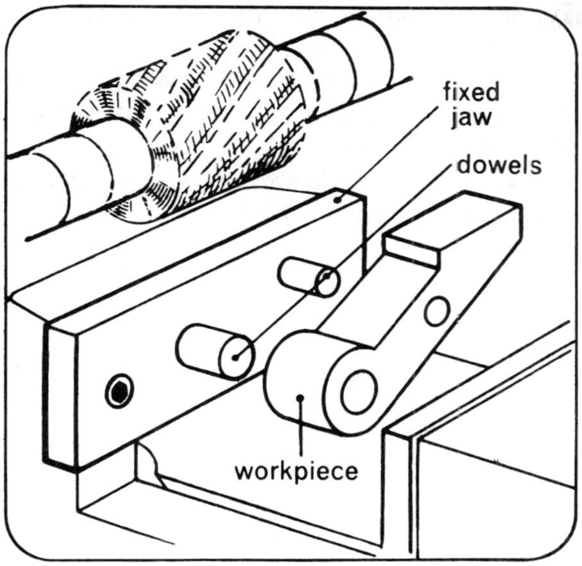

4 Setting fixture for prebored workpiece

(a) The fixed jaw is bored and fitted with dowels to suit prebored workpiece. Dowels must be shorter than thickness of workpiece.

(b) Position workpiece, locating on dowels.

(c) Close vise to secure workpiece, using flat surface moving jaw.

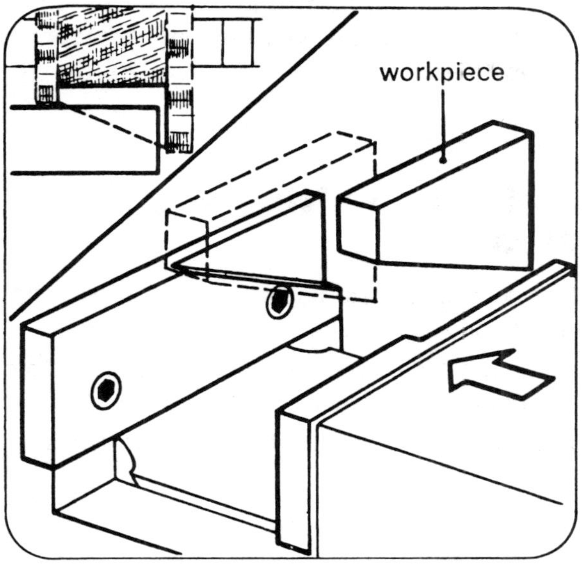

5 Setting fixture using angular faces

(a) Prepare soft jaws by milling angles on each jaw, deep enough to locate workpiece but with space to secure by vise pressure.

(b) Insert workpiece in required position and tighten vise.

(c) Mill workpiece up to second operation.

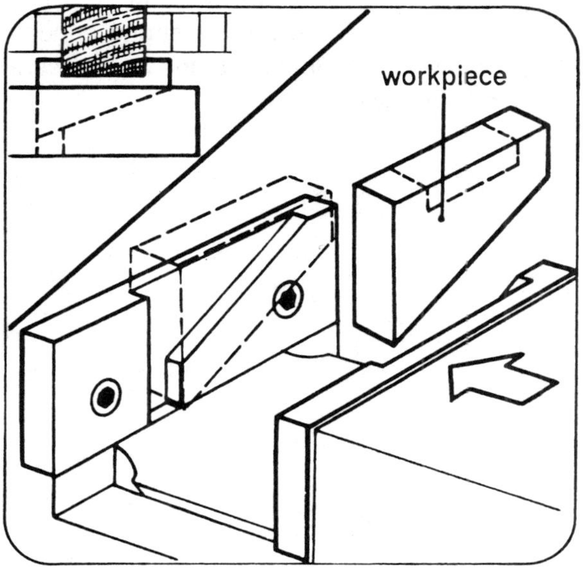

(d) Prepare soft jaws to complete second operation.

Note:
It is desirable to ensure clearance of corners for setting purposes and to allow chip clearance.

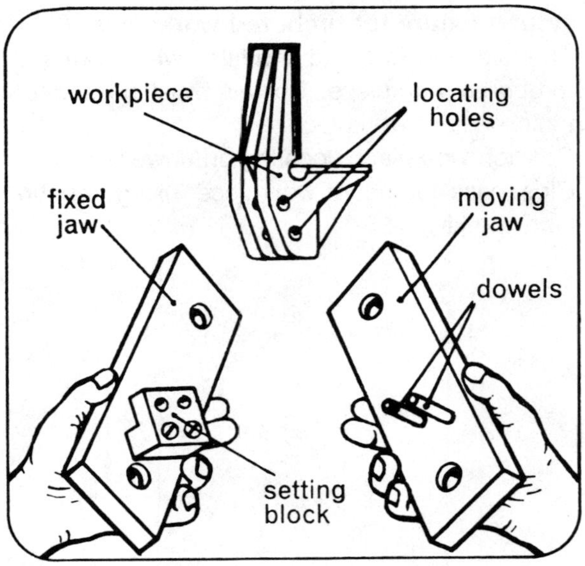

6 Combined fixture for partially machined workpiece

(a) Prepare fixed soft jaw with setting block to locate slot in base of workpiece and holes in setting block bored to dimensions required.

(b) Prepare moving soft jaw with dowels inserted to correspond with holes in setting block.

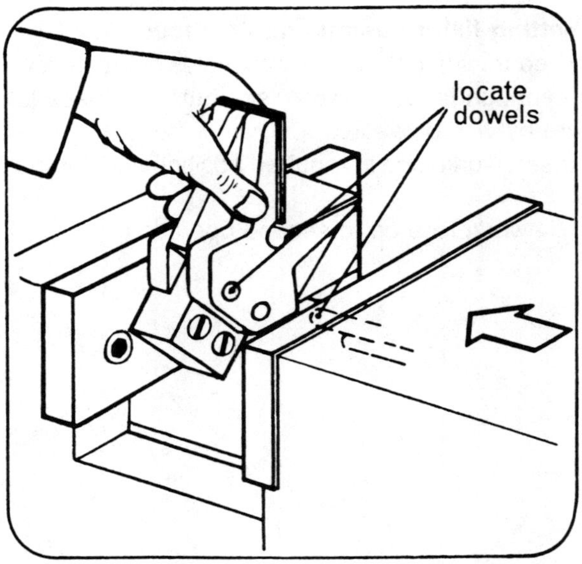

(c) Insert workpiece, holding in approximate position, and close vise, ensuring that dowels properly locate workpiece.

Slinging and lifting angle plates

1 Decide approximate position for angle plate on table after checking drawing.

2 Clean machine table and angle plate. Stone off any burrs on plate or table.

3 Place bolts in table slots in desired positions.

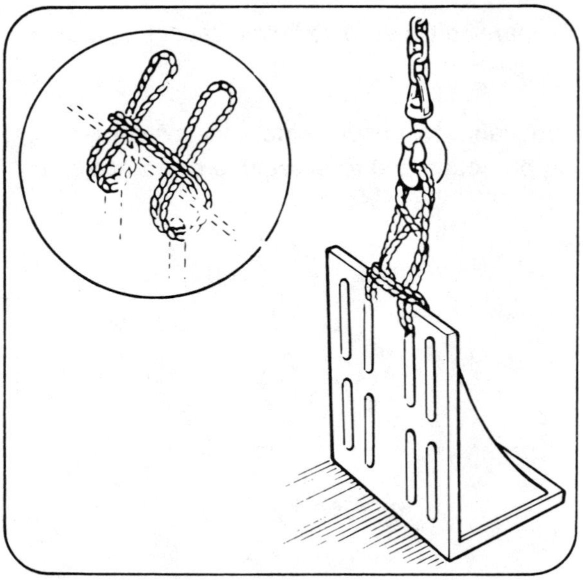

4 Sling angle plate as shown in diagram. Sling checks to be sure that the rope is strong enough and free from wear, and that the crane hook is centered. Stand clear when lifting.

5 Lower angle plate gently onto machine and bolt in position.

> **Safety**
>
> Ensure that the crane hook is clear of aisles and work area after use.

Setting an angle plate parallel

1 Set angle plate parallel

(a) Place a pair of keys in table slot, ensuring that they are a good fit.
(b) Place pieces of paper, or two equal parallels, between keys and face of angle plate.
(c) Push angle plate hard toward keys and partially tighten clamping bolts.

> **Safety**
>
> When moving the angle plate on the machine table take care not to overbalance the plate and let it fall from the table.

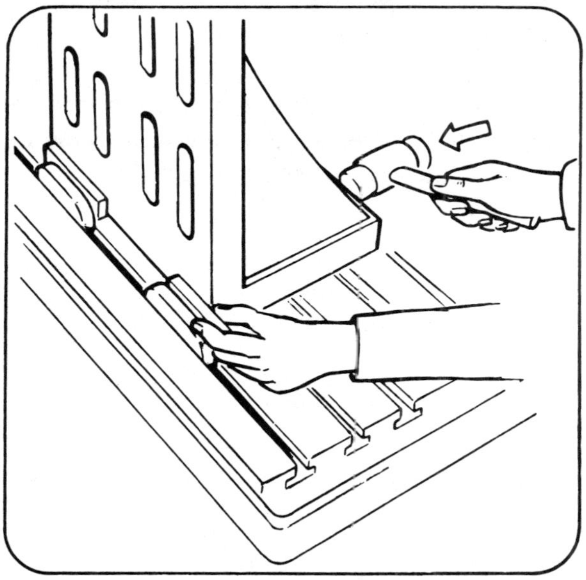

2 Check and adjust setting

(a) Using soft hammer, tap base of angle plate toward key at end where movement, if any, is felt.
(b) When both papers or parallels are trapped, tighten clamping bolts securely.

Setting an angle plate square

1 Set angle plate approximately square on table

2 Position dial indicator

(a) Attach magnetic base of dial indicator to vertical slide.

(b) Place stylus of dial indicator square to faceplate and clear of slots. Lock all indicator joints.

3 Check alignment

(a) Lock one clamping bolt and traverse the table to bring dial indicator to fixed end. Note reading.
(b) Adjust faceplate by tapping with soft hammer until a zero reading on dial indicator can be maintained across angle plate when table is traversed.
(c) Secure all bolts. Recheck alignment.

Setting an angle plate face vertical

The techniques are similar to that used for setting an angle plate square, but the vertical traverse of the table is used instead (see diagram).

Adjustments are made using shims as follows:

1 Establish the error in the vertical plane, and note the direction for adjustment.

2 Calculate the thickness of the shims required as follows:

$$\text{shim thickness} = \text{error} \times \frac{\text{base length of faceplate}}{\text{vertical length of faceplate}}$$

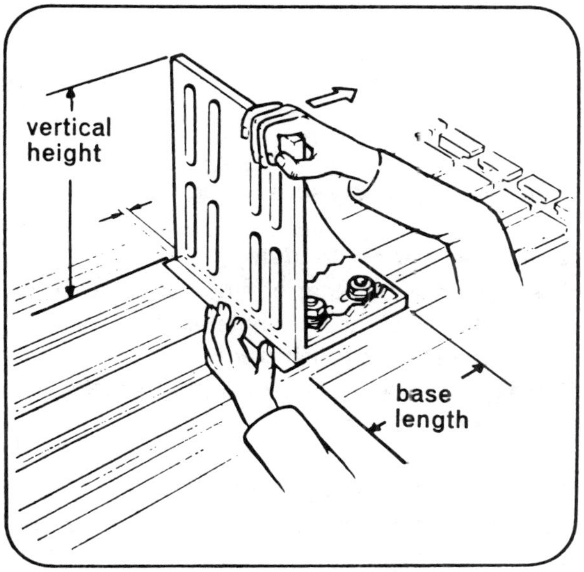

3 Place the shims at the front or rear *edge* of the faceplate as required. Tighten facebolts and recheck alignment.

Setting an angle plate at an angle

1 Set angle plate at approximate angle

(a) Swivel angle plate, visually judging angle of face to machine table.

(b) Lightly tighten one bolt to form pivot point.

2 Check angular setting

(a) Set protractor at required angle, ensuring that blade does not project beyond protractor base.

(b) Hold protractor base firmly on vertical slide with blade just above, and parallel to, machine table. Ensure that maximum length of blade reaches across face of angle plate.

(c) Slide protractor toward face of angle plate until edge of blade touches highest point.

(d) Look for any gap between protractor blade and face of angle plate.

3 Tap angle plate to angular setting

(a) Tap face of angle plate away from protractor at end touching blade.

(b) Maintain pressure on protractor to hold base on vertical slide and blade up to face.

(c) Tap until no gap is visible.

4 Accurately set angle plate

(a) Position piece of paper or strips of equal thickness at each end of face of angle plate.

(b) Slide protractor to touch blade to paper.

(c) Tap face of angle plate until both papers are tightly trapped by blade.

5 Final clamping of angle plate

(a) Take care when clamping to avoid moving angle plate.

(b) Check angular setting after clamping.

Setting an angle plate using the sine bar

1 Check horizontal table alignment

2 Mount plate

(a) Locate two setting blocks in table slot at distance to suit length of protractor base.

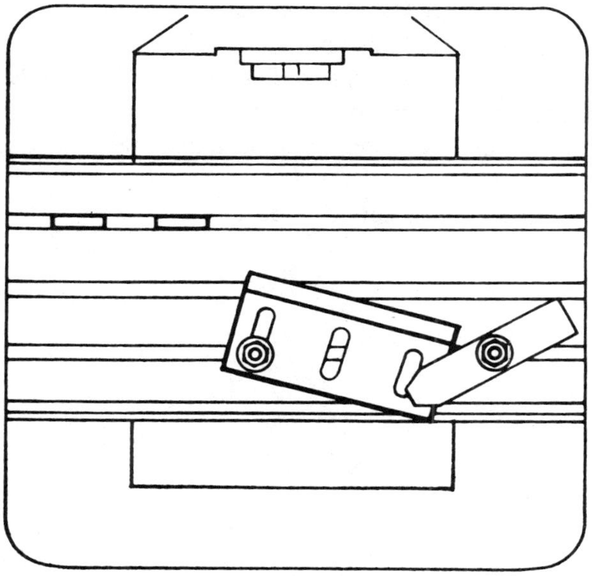

(b) Mount angle plate visually to position desired. Lightly bolt or clamp.

3 Set at approximate angle

(a) Set protractor at required angle.

(b) Hold protractor base firmly in contact with blocks and, with blade parallel to table, check angle face of plate.

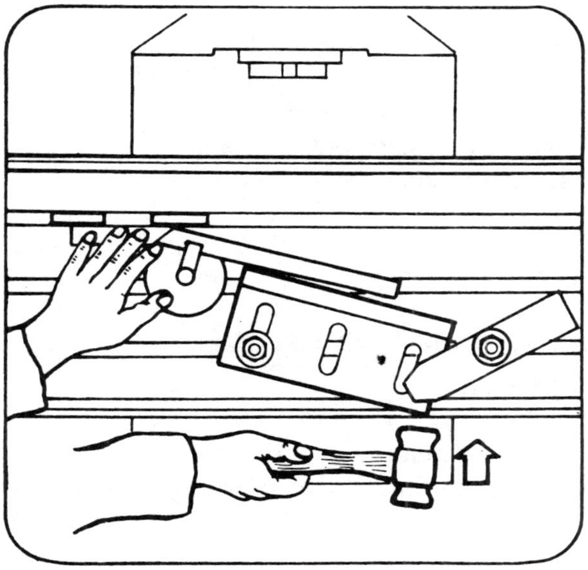

(c) Adjust angle plate by tapping with soft hammer until there is no gap between plate and protractor blade.

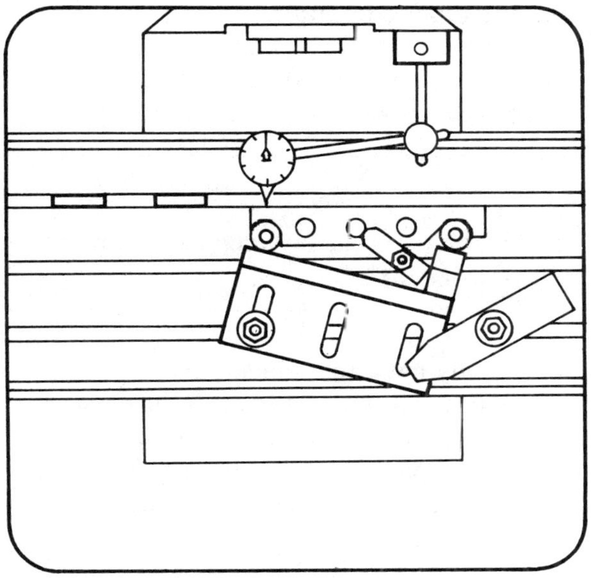

4 Position sine bar

(a) Select and clean appropriate size sine bar.

(b) Calculate, clean, and wring together gage blocks to set sine bar at required angle.

(c) Locate sine bar with rolls and gage blocks in contact with angle block faces.

(d) Very lightly clamp, using protective pad between clamp and sine bar.

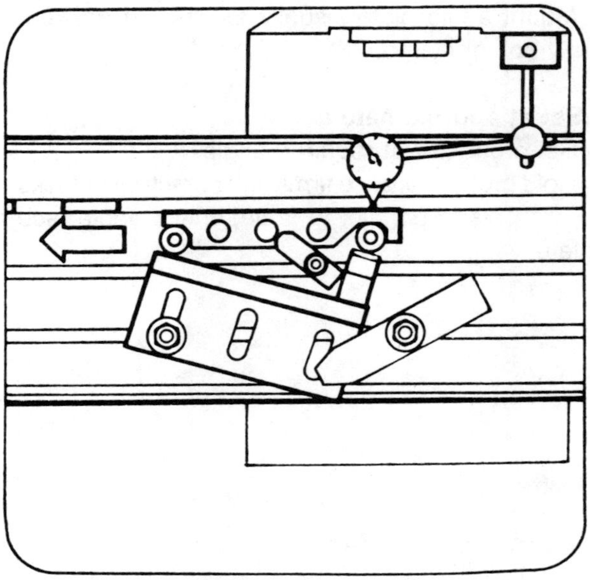

5 **Position and set dial indicator**

(a) Attach dial indicator base to vertical slide.

(b) Locate and load dial indicator stylus and set bezel to zero on face of sine bar at one end.

(c) Move table to bring stylus to other end of sine bar face.

(d) Note any deviation from zero.

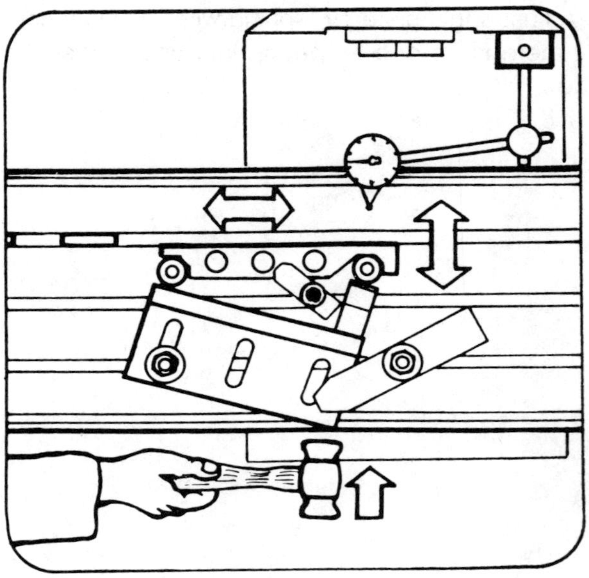

6 **Adjust**

(a) Release sine bar from contact faces.

(b) Adjust angle plate by tapping with soft hammer.

(c) Reposition and clamp sine bar.

(d) Recheck as steps 5(b) and 5(c).

(e) Tighten bolts or clamps.

(f) Recheck.

Setting an adjustable angle plate

1 **Set angle plate**

Set angle plate square or parallel unless compound angle is required.

2 **Tilt angle plate to angle**

(a) Loosen swivel retaining bolts.

(b) Tilt angle plate to angle required.

(c) Tighten swivel retaining bolts lightly.

> **Safety**
>
> Be careful of your hands when loosening and tilting the angle plate.

3 **Accurately set angle plate**

(a) Set protractor at required angle. Ensure that blade does not extend beyond protractor base.

(b) Position piece of paper or shims of equal thickness at top and bottom of angular face.

(c) Hold base of protractor firmly onto machine table. If necessary raise protractor on parallel(s) to lift blade to height of angular face.
(d) Slide protractor to trap paper with blade, visually checking that blade is parallel to end of angle plate.
(e) Pull papers lightly to feel if loose.
(f) Tap face of angle plate, at top, up or down, until both papers are tightly trapped.

4 Final tighten
(a) Tighten swivel retaining bolts securely.
(b) Check that angular setting has not moved.
(c) Reset correct angle if movement has occurred.

Workholding in chucks

1 Position sling
(a) Lift one side of chuck and stand on edge.
(b) Sling chuck as shown in diagram. Sling checks to be sure that rope is strong enough and free from wear and that crane hook is centered. Stand clear when lifting.
(c) Clean chuck base. Stone off any burrs on base or table.

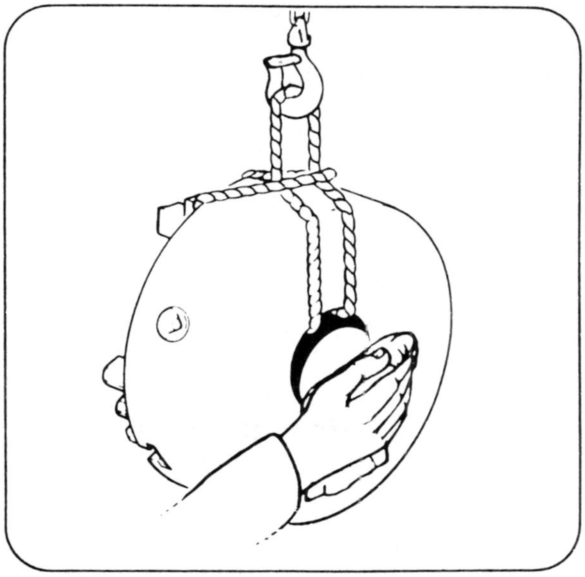

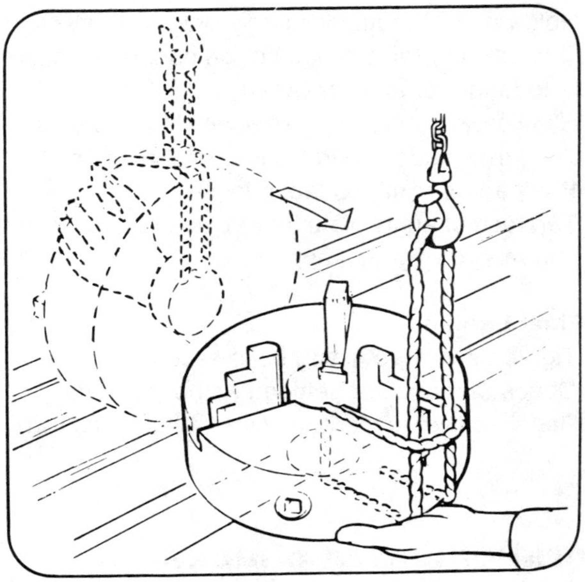

2 Position chuck on table

(a) Lower chuck gently onto table, positioning base edge in table slot at far side of bed.

(b) Pull top of chuck toward body. Lower crane steadily until base of chuck is seated on table.

(c) Slide chuck as far on table as possible and support front of chuck. Remove sling from hook and from bore of chuck.

(d) Slide chuck to center of table with chuck key socket toward front of machine.

> **Safety**
>
> Do not allow the chuck to overbalance or fall from the machine table.

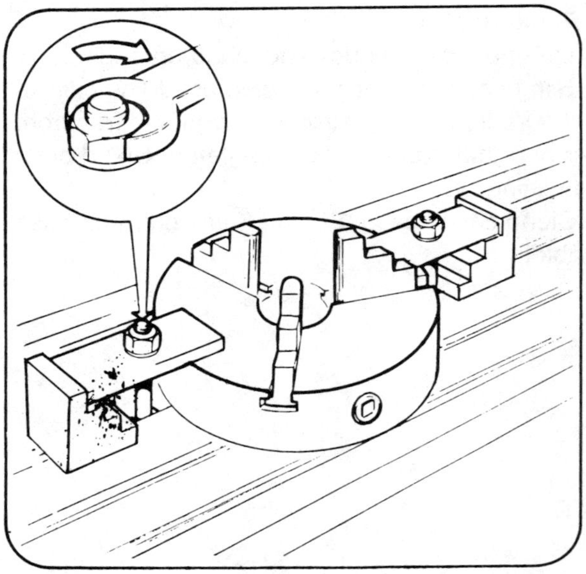

3 Clamp chuck

(a) Position table bolts and clamps one on each side of chuck.

(b) Ensure that clamps do not hit chuck jaws.

(c) Ensure that bolts will not hit cutter.

(d) Tighten clamps securely.

Setting and aligning vee blocks

1 Select vee blocks

(a) Select a pair of vee blocks that will allow diameter of shaft to rest approximately midway down groove faces.

(b) Ensure that vee blocks are clean and free from burrs. Stone off any burrs.

2 Position vee blocks

(a) Refer to drawing for machining operation(s) and position workpiece and vee blocks on machine table.

(b) Ensure that blocks will clear cutters.

(c) Position two table bolts in rear table slot and ensure that length of bolts is suitable for clamping on top of shaft.

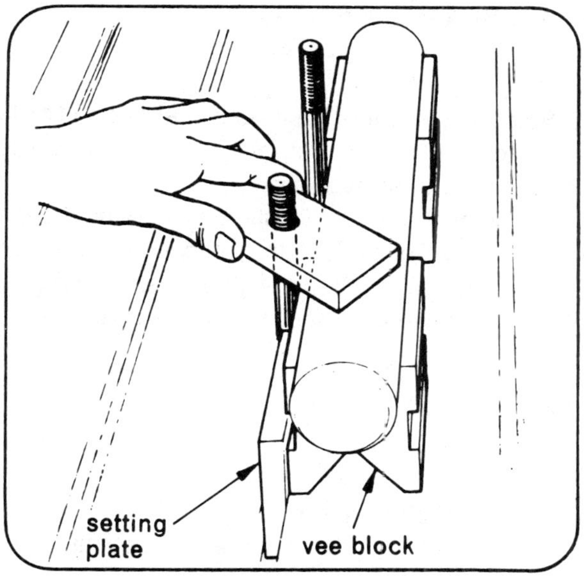

(d) Position two keys or setting plates in rear table slot, ensuring that they are a sliding fit with no play and will not obstruct clamps on top of shaft.

(e) Push vee block up to setting plate and check that base of block is located clear of table slots. Parallels may be inserted between vee blocks and setting plates if necessary.

3 Clamp vee blocks

(a) Position clamps on top edges of vee blocks, ensuring that they do not hit shaft.

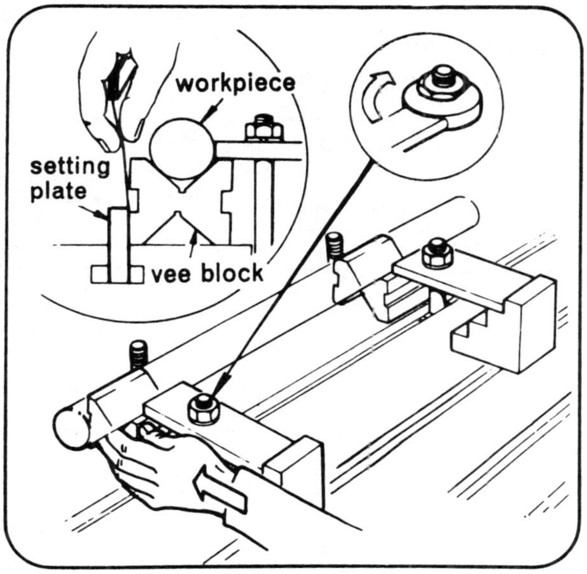

(b) Push vee blocks hard up to setting plates and tighten clamps securely, ensuring that blocks do not tilt over.

(c) Check with feeler gage that vee blocks remain tight up to setting plates. Alternatively, tightness may be checked by trapping a piece of paper before clamping and ensuring that it remains well trapped.

4 Clamp workpiece in vee block

(a) Position clamps on workpiece, central to and directly over, vee blocks.

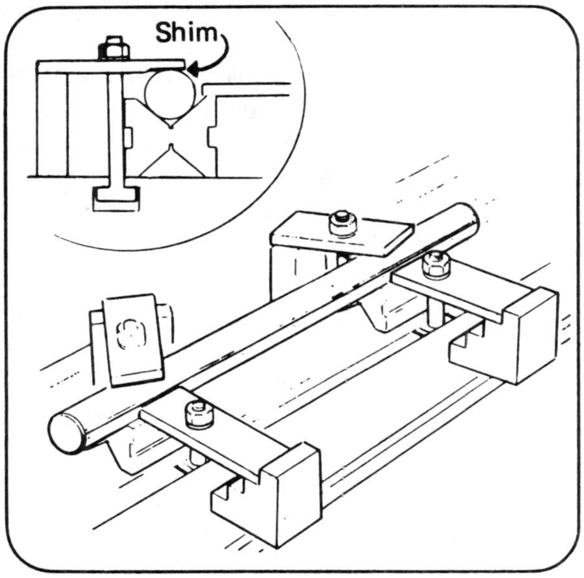

(b) Protect finished diameters by inserting piece of soft material between clamp and workpiece (e.g., a piece of abrasive cloth folded to about four thicknesses).

(c) Ensure that clamps will not hit cutters.

(d) Tighten clamps securely.

Rotary tables

1 Sling and lift

(a) Select two suitable T nuts and eyebolts and slide them into center table slot, approximately 1 in. (25 mm) from edge, at each side.

(b) Screw eyebolts into T nuts, ensuring that flange of bolt is seated on table. Sling rotary table as shown in diagram. Check that rope or chain is strong enough and free from wear. Crane hook must be centered.

2 Set on machine table

(a) Clean base of rotary table and machine table. Stone off burrs.

(b) Lower rotary table gently onto machine table with handle facing front.

(c) Lightly clamp to machine table.

> **Safety**
>
> Stand well clear when lifting the rotary table on the crane.

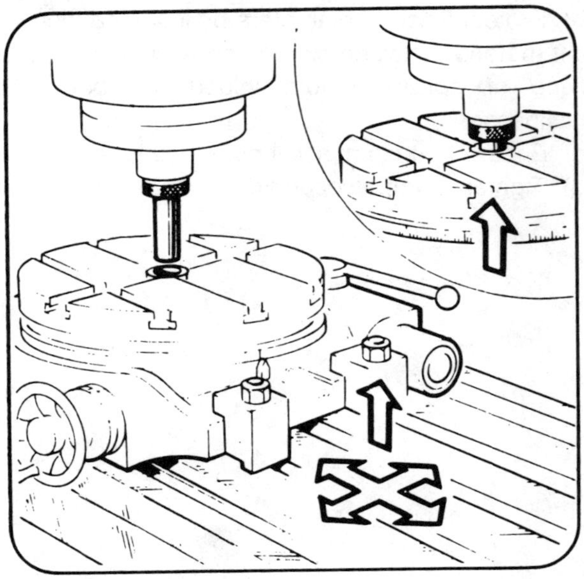

3 Align vertical head spindle and rotary table axis

(a) Select and mount a suitable taper shank setting bar. (Diameter of slide fit to bushing.)

(b) Clean bushing in center bore of rotary table.

(c) Ensure that table bolts are hand tight only.

(d) Raise and position table to allow bar to slide into bushing.

4 Set attachment relative to table movement

(a) Zero both indexes and lock both slides.
(b) Tighten table bolts securely.
(c) Disengage setting bar.

Alternatively:

Bolt the attachment securely to the machine table. Mount the setting plug in the table bushing. The central position is obtained by using a dial indicator mounted on the arbor. Ensure that the direction of the table setting is marked after final setting.

Mounting and clamping workpiece to rotary table

1 Mounting the workpiece

(a) Clean table and zero scale.
(b) Select and mount suitable parallels.
(c) Center workpiece approximately on table axis.

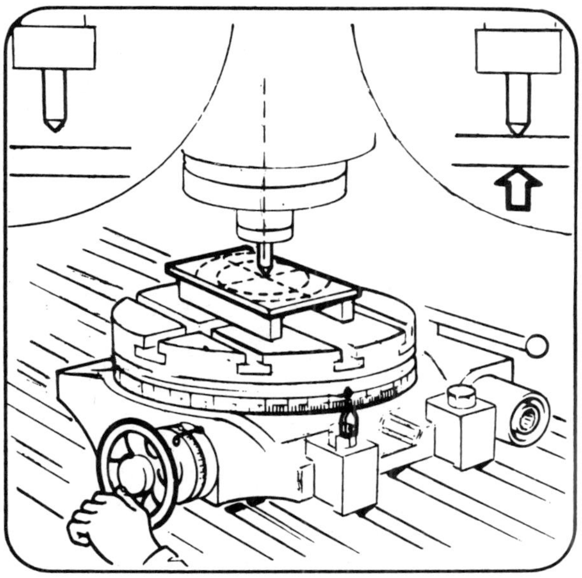

(d) Mount a center in machine spindle, visually checking that center point of workpiece and center in machine are in line. Engage center lightly on workpiece, ensuring that centers have met.

Note:

If holes are used for centering the workpiece, these should be drilled and reamed and a plug fitted in the hole. Use the dial indicator technique for accurate alignment.

2 Mounting workpieces without marked centers

(a) Set attachment relative to table movement.
(b) Move table required distance from centerline, for radius to be machined.
(c) Ensure that final setting is indexed in appropriate direction.

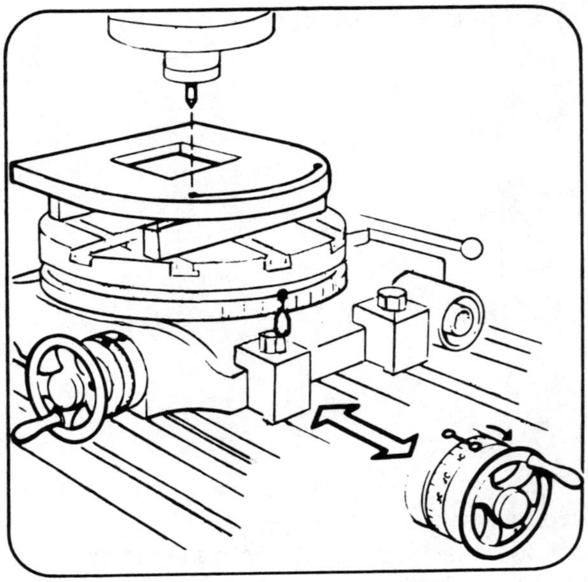

(d) Zero index.

(e) Mount center in machine spindle, and position workpiece to align center with marked radius line.

(f) Hand feed table around, gently sliding workpiece to position so that center follows marked radius line.

(g) Attach clamp assembly, and tighten fully when radius line is in exact position.

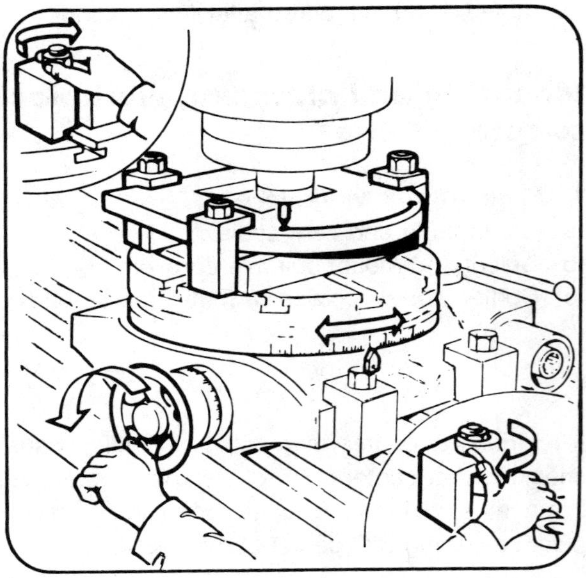

3 Note on clamping the workpiece

(a) Place clamps so that if possible machining operation can be completed without cutter hitting clamps.

(b) Ensure that clamps are placed directly over parallels under workpiece. It is not always possible to mill some types of workpiece without moving a clamp. Always ensure that one clamp is moved at a time, and that other clamps are kept tight to ensure that workpiece does not move.

Review questions

1 Give some general tips on workholding and setting.

2 What safety precautions should be observed when slinging and lifting heavy mill accessories?

3 Describe and demonstrate vise alignment on the milling machine.

4 How are mill vises set at angles?

5 How would you align two vises on the same machine?

6 How are swivel and universal vises set at angles?

7 What are soft jaws, and how are they used?

8 How may mill vise jaws be configured for production machining?

9 How are angle plates aligned?

10 How is the face of an angle plate checked and adjusted for vertical alignment?

11 How is an angle plate set at an angle?

12 How is the sine bar used in angle plate setup?

13 How might a chuck be used for mill workholding?

14 How are vee blocks used and aligned?

15 Describe alignment and workholding on a rotary table.

UNIT SEVEN

Setting the Workpiece

Clamping directly to the machine table

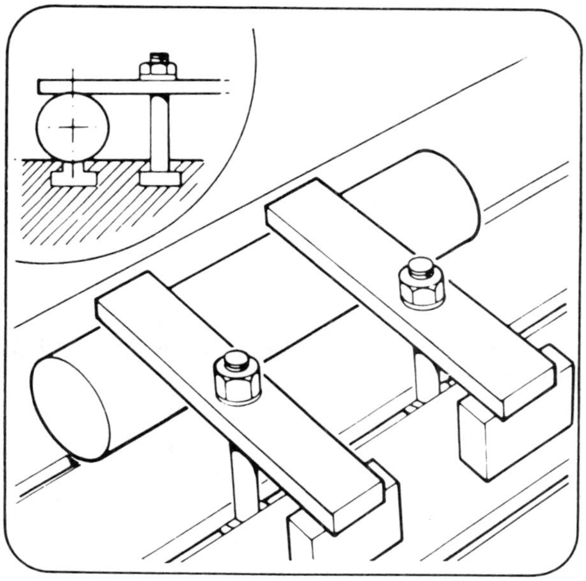

1 Use of table slots for locating shafts

(a) Check that top corners of table slot are free of burrs. Stone if necessary.

(b) Locate shaft in table slot.

(c) Clamp shaft directly over slot.

2 Use of table slots for aligning workpiece

(a) Position workpiece centrally on table.

(b) Engage keys or setting plates in table slot, within length of, but as near ends of workpiece as possible.

(c) Check that keys are a good fit in table slot with no side play.

(d) Push workpiece to keys or setting plates and visually check that base is seated on table clear of slots. Parallels may be inserted between keys and workpiece if required.

(e) Hold workpiece against keys while clamping.

(f) Check with feeler gage that workpiece is tight against keys. Alternatively, ensure that paper strips between workpiece and keys remain tightly trapped.

3 Use of table slots for setting workpiece square or at angle

(a) Position setting plates as close as possible to workpiece. Ensure good fit.

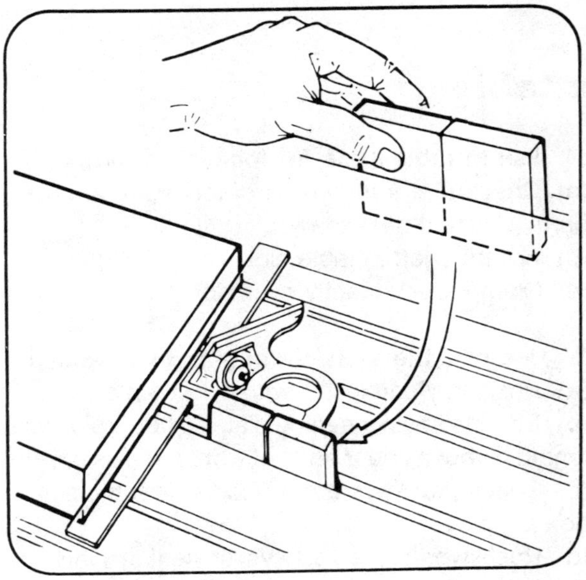

(b) Ensure that maximum possible amount of square or protractor base is against setting plates.

(c) Hold base flat up to setting plates and slide blade up to face of workpiece.

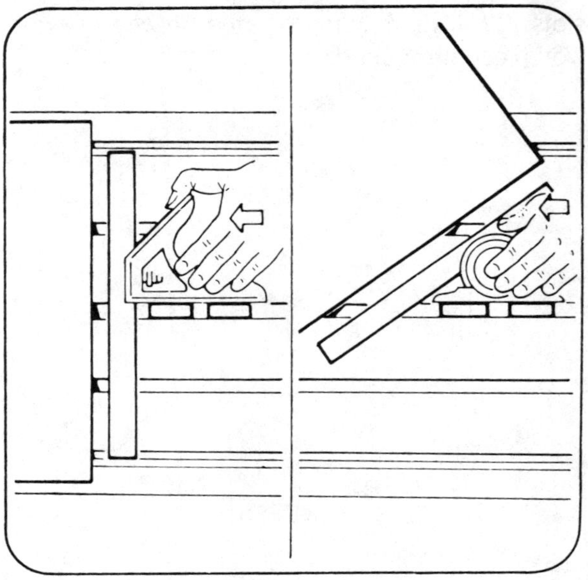

(d) Tap workpiece to align face to square or protractor blade.

Clamping regular shaped work in a vise

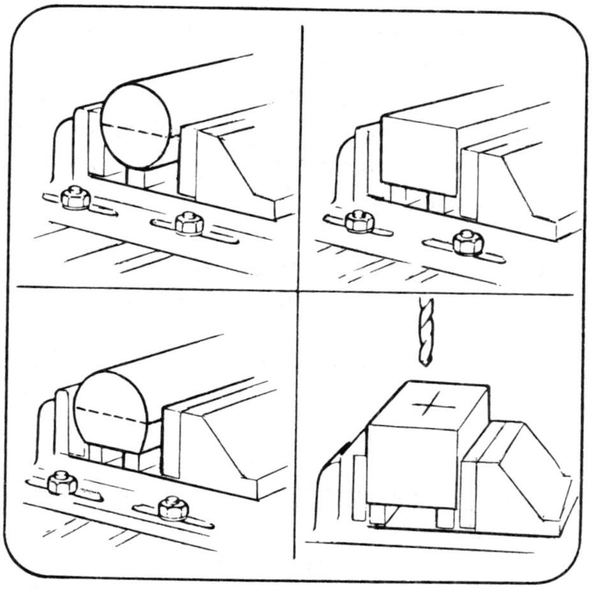

1 Position workpiece in vise

(a) Position maximum possible length and depth of workpiece in vise, ensuring that cutters will not hit vise.

(b) If necessary to raise height of workpiece, position one parallel centered (if bottom of workpiece is round) or one parallel to each edge (if bottom of workpiece is flat).

(c) Check that parallel will not hit cutters or measuring instruments.

(d) Visually check that centerline of round workpiece is below top of vise jaw.

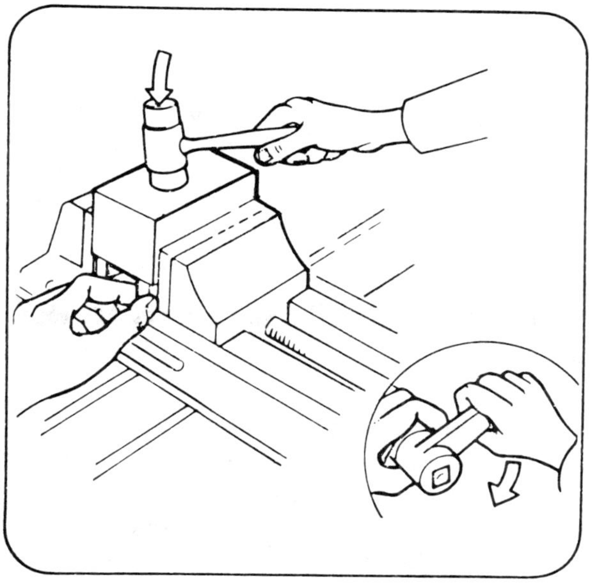

2 Seat workpiece and tighten vise

(a) Lightly tighten vise and tap workpiece down with soft hammer. Ringing sound is heard until workpiece is seated; then sound is dull and thudding.

(b) Check with feeler gage that workpiece is correctly seated to vise bed. If parallels are used, ensure that they are tightly trapped.

(c) Tighten vise.

Clamping irregular shaped work in a vise

1 Position and secure workpiece

(a) Use a rack vise.

(b) Position workpiece approximately centered on vise jaws with largest clamping face to fixed jaw of vise.

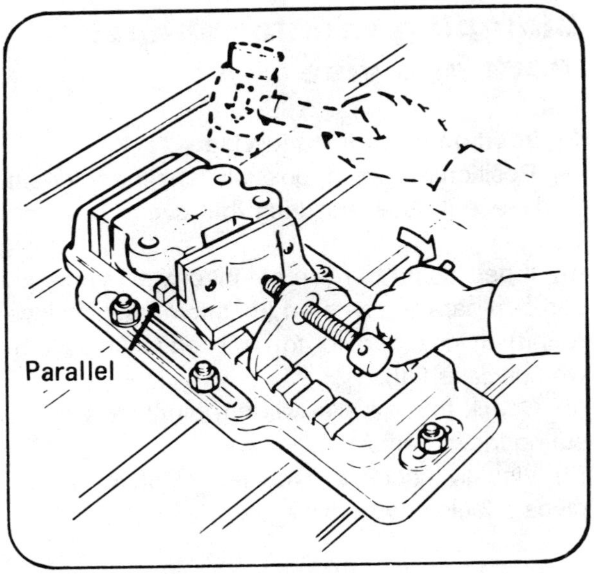

(c) Slide movable jaw up to workpiece swiveling vise jaw to obtain maximum clamping surface.

(d) Position screw holder as close as possible to rear of movable jaw.

(e) Position parallels as required to raise level of finished face above top of vise jaw.

(f) Tighten vise jaws lightly and tap workpiece level.

(g) Tighten vise jaws securely, but do not over-tighten because this may result in distortion or breaking of workpiece.

2 Release stress for finish cut

(a) Loosen vise jaws until only light resistance to turn of screw is felt.

(b) Retighten vise as tightly as possible without handle.

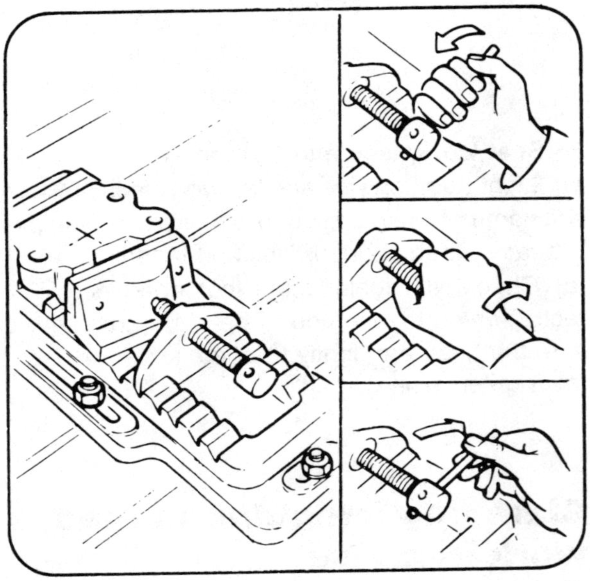

(c) Engage vise handle and tighten vise very lightly with hand pressure only.

(d) Check setting of workpiece.

(e) Take only light cut(s) after releasing stress.

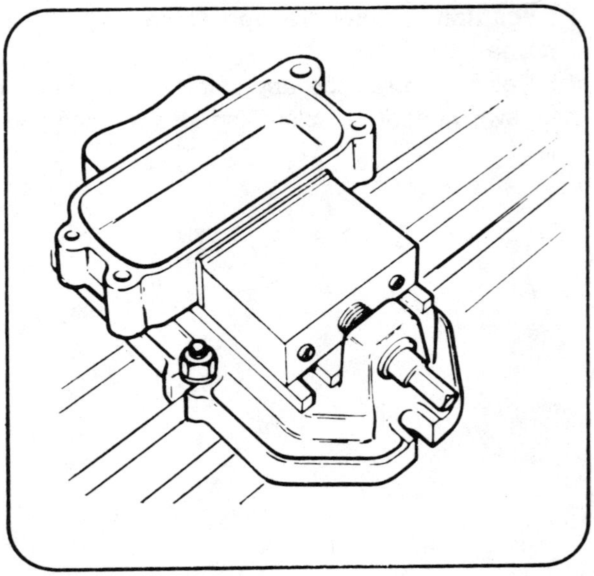

Clamping hollow work in a vise

1 Position workpiece in vise

(a) Position workpiece, ensuring that maximum length possible is between vise jaws.

(b) Position shims or parallels, if required, to raise level of face to be machined just above top of vise jaws and to set workpiece level.

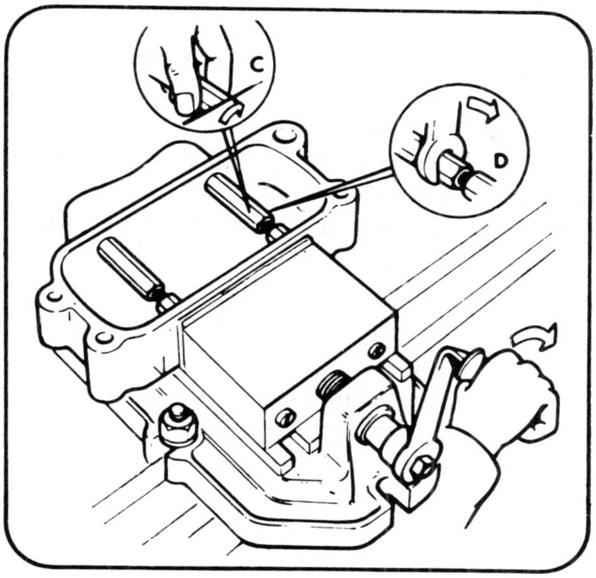

2 Clamping workpiece in vise

(a) Tighten vise jaws lightly on workpiece.

(b) Position small jacks inside workpiece, ensuring that they are within length of and just below top of vise jaw.

(c) Finger tighten jacks to walls of workpiece.

(d) Engage wrench and lightly tighten jacks, allowing only hand pressure to turn nut.

(e) Tighten vise with handle as tight as possible by hand.

3 Release stress for finish cut

(a) Loosen vise jaws until moderate resistance to turn of screw is felt.

(b) Loosen jacks to finger tight.

(c) Loosen vise until only light resistance to turn of screw is felt.

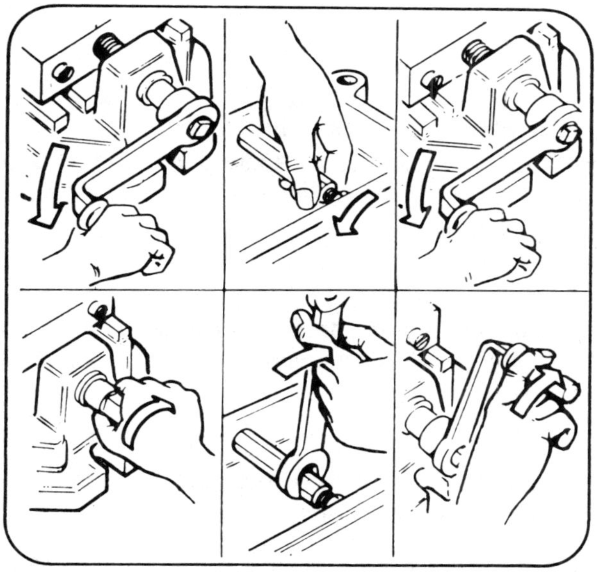

(d) Retighten vise as tight as possible without using handle.

(e) Finger tighten jacks.

(f) Engage wrench and tighten jacks applying light pressure with fingers.

(g) Tighten vise with hand pressure only on handle.

(h) Check setting of workpiece.

(i) Take only light cut(s) after releasing stress.

Note:

Stress is released for final machining only. Failure to release stress will result in distorted faces, bores, and so on when the workpiece is removed from the vise.

Clamping work to an angle plate

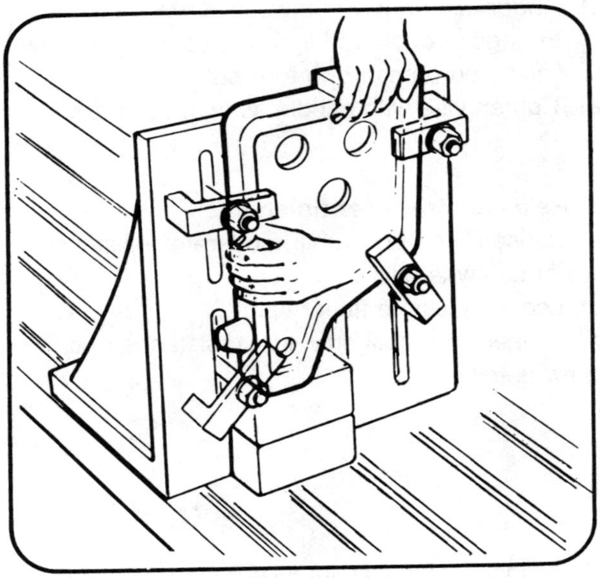

1 Mount angle plate on machine table

2 Preposition clamps

(a) Hold workpiece to angle plate. Select appropriate slots to give maximum clamping area.
(b) Position parallels, shims, or jacks to support workpiece in correct position.
(c) Remove workpiece and position table bolts, or T nuts and studs, through slots of angle plate with stem of T in slots.
(d) Build up clamps on angle plate and place clamp blocks. Keep wrench nearby ready for use.

3 Position and clamp workpiece

(a) Position workpiece to angle plate, and hold in position with C clamp, leaving hands free for positioning clamps.
(b) Position and hold clamp to workpiece, block up rear of clamp, and finger tighten nut.
(c) Position other clamp(s) and tighten bolts.
(d) Check setting of workpiece and adjust if necessary.
(e) Tighten clamps.

For a workpiece without a premachined face, use the same procedure except that:

(i) Blocking must be positioned to align workpiece and give support under clamps.
(ii) Stress must be released before taking finish cut.

4 Releasing clamping stresses

(a) Loosen one clamp to finger tight, ensuring that workpiece does not tilt from clamping.
(b) Tighten clamp lightly with hand pressure only on wrench.
(c) Repeat steps (a) and (b) for other clamps.

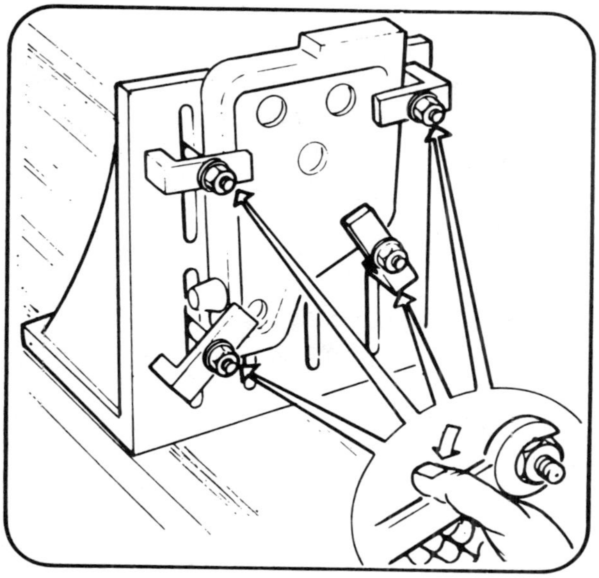

(d) Check that even pressure is applied on all clamps.
(e) Check setting of workpiece.
(f) Take only light cut(s) after releasing stress.

Note:
Clamping pressures are released for final machining when there is a danger of the workpiece being distorted by clamping pressures.

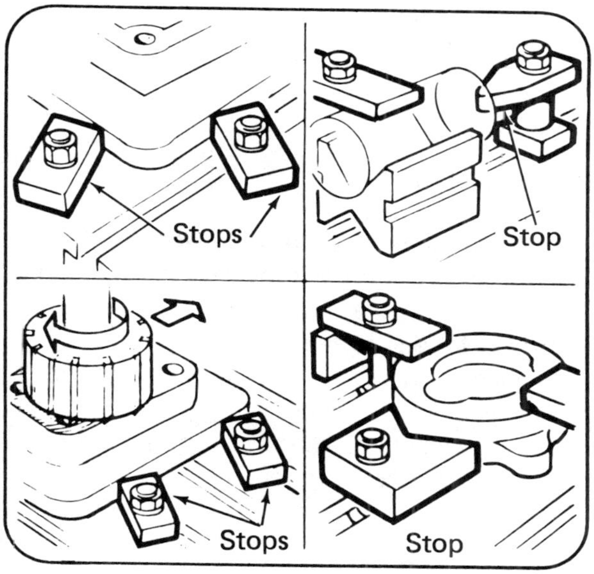

Using work stops

If several workpieces are to be machined, stops can be used to locate them in the same position on the table. This avoids setting each item individually.

Stops also counteract the thrust of the cutters, which otherwise might move the workpiece.

1 Correctly set workpiece

2 Set stops
(a) Position stops to sides and/or end of workpiece to control position in required plane(s).
(b) Ensure that workpiece is clean where stops are positioned.

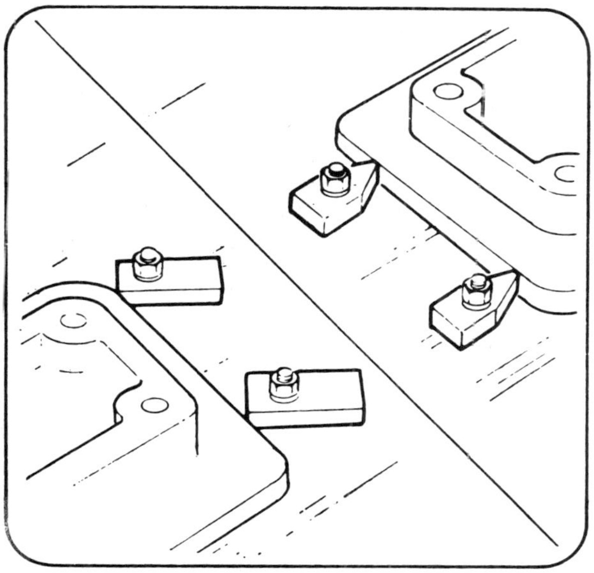

(c) Position stops so that only small area is in contact with workpiece. Use clamp with tapered sides, or slightly turn clamp to locate on corner.

3 Clamp stops

(a) Clamp as near locating end of stop as possible.
(b) Ensure that stops and bolts will be clear of cutters.
(c) Hold stops in position and tighten bolts.
(d) Tap stops until slight vibration is felt through workpiece, indicating that stops are located against workpiece.
(e) Ensure that cutter thrust is toward stops.

4 Position workpiece to stops

(a) Clean locating points of stops and workpiece before positioning.
(b) Hold workpiece tight up to stops and clamp in position.

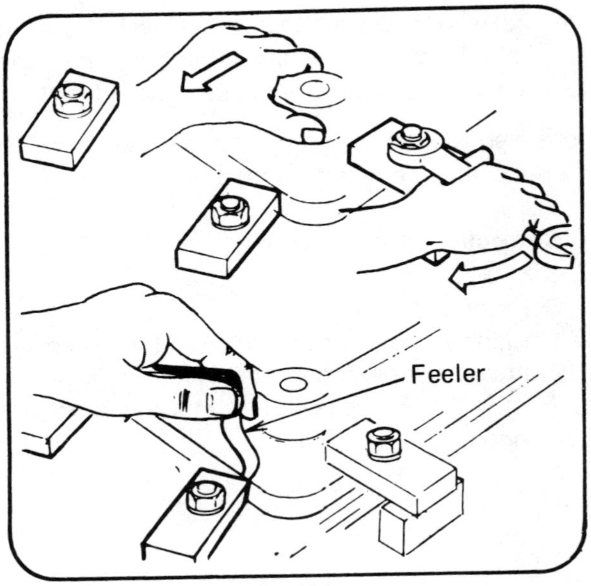

(c) Check location of workpiece with feeler gage or trapped paper.

Edgefinding

1 Set workpiece and clamp in position

Accurately align face that is to be used as reference.

2 Set workpiece to face

(a) Engage setting bar in machine spindle and check concentricity.
(b) Roughly center setting bar on face, approximately .010 in. (.25 mm) from it.

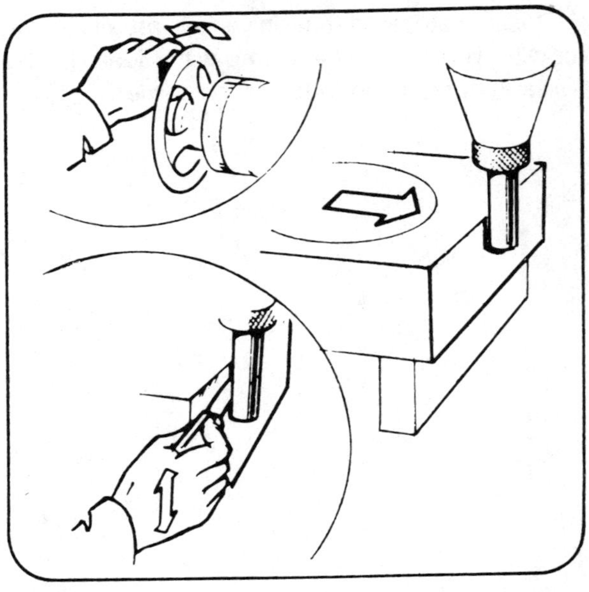

(c) Position feeler gage between setting bar and face.
(d) Slide feeler gage up and down and move table toward bar until feeler gage is lightly trapped but still can be moved up and down.

3 Set centerline of spindle to reference face

(a) Lower table to allow bottom of setting bar to clear workpiece.

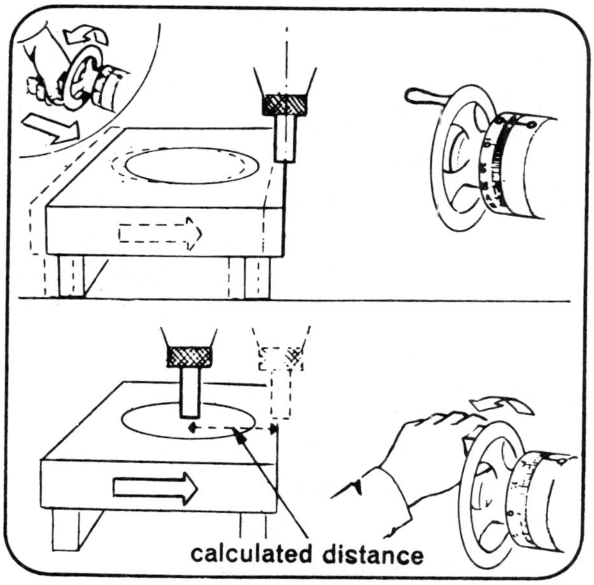

(b) Hand crank table, half the diameter of setting bar plus thickness of feeler gage used, to bring centerline of setting bar directly over reference face.
(c) Zero index.

The table may now be traversed to the required position.

Note:
If setting is required in two planes, the same procedure is used.

4 Indexing to position spindle
Calculate amount of movement required, and move table to required position.

5 Indexing to face to be machined
(a) Set cutter in machine spindle and measure its diameter.
(b) Add half the diameter of cutter to given dimension if face(s) are on opposite side of workpiece from reference.
(c) Subtract half the diameter of cutter from given dimension if face(s) are on same side of workpiece as reference.
(d) Calculate amount of table movement required, and move the table the calculated amount.
(e) Accommodate backlash where necessary.

Using the offset edgefinder

1 Clamp and align the workpiece.

2 Install an offset edgefinder in the chuck or collet.

3 Start the machine spindle at moderate RPM.

4 Gently move the workpiece against the edgefinder tip and observe tip movement.

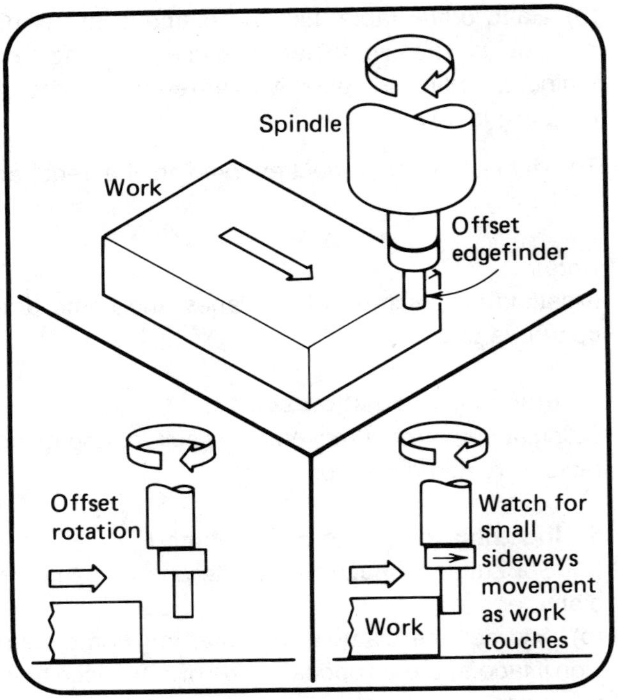

5 Continue to move the work gently until the edgefinder tip springs sideways.

6 Stop movement and set zero on the table or saddle micrometer collar.

7 The workpiece edge is now located half the edgefinder tip diameter from the spindle centerline.

8 Remove the edgefinder or raise it above the workpiece. Move the table an additional distance equal to the edgefinder tip radius (usually .100 in.).

Using a wiggler center finder

1 Clamp and align the workpiece.

2 Install a wiggler in the chuck or collet.

3 Start the machine spindle at fast RPM.

4 Use a pusher to touch the rotating wiggler tip. Push gently toward the center until the wiggler tip runs true.

Safety

Keep your hands and fingers clear of the wiggler tip.

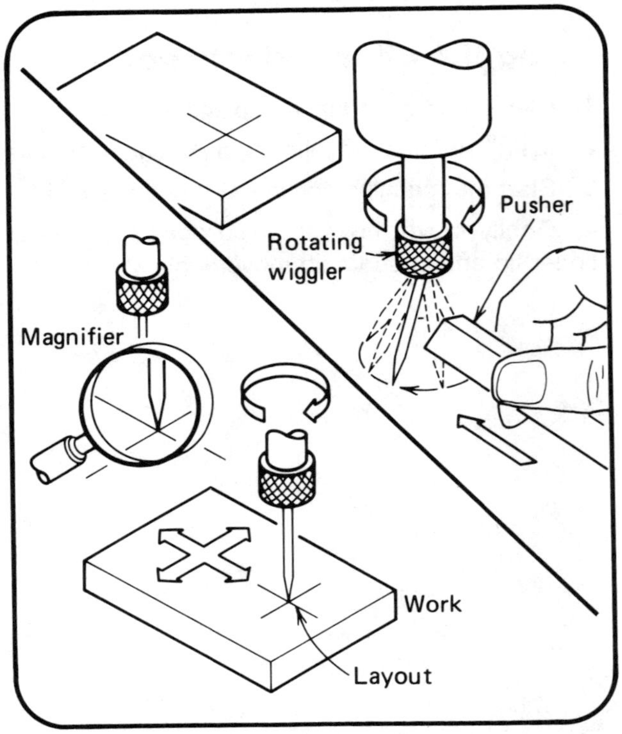

5 Position the tip just above the workpiece layout, but do not let the tip touch the workpiece.

6 Move the table and saddle until the workpiece layout is centered under the wiggler tip. Use a magnifying glass for more accurate locating.

Using the centerscope

1 Clamp and align the workpiece.

2 Install the centerscope in a chuck or collet.

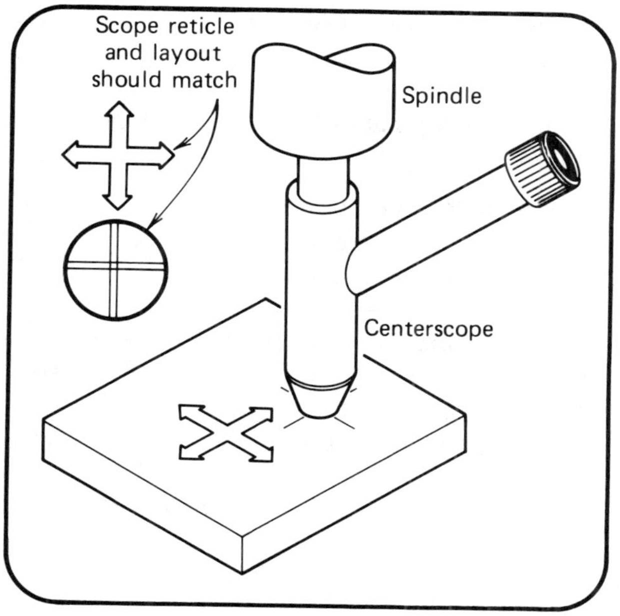

3 Move the workpiece edge or layout until it is aligned to the centerscope cross hair.

4 Zero the table and saddle micrometer collars.

Positioning centerline of spindle to centerline of horizontal shaft

Methods:

1 Touching cutter to thin paper attached to side of shaft.

2 Touching cutter to layout dye or chalk on side of shaft.

3 Setting with feeler gage between cutter and shaft.

4 Touching setting bar to side of shaft.

5 Setting with feeler gage between setting bar and shaft.

6 Using dial indicator in machine spindle.

Method 1 Touching cutter to thin paper attached to side of shaft.

1 Clamp shaft in position
(a) Select appropriate workholding method.
(b) Correct alignment of shaft is important.

2 Place paper on shaft and position cutter
(a) Use tissue paper or similar very thin paper. Moisten paper only enough to attach to shaft.
(b) Position cutter in line with paper on shaft.

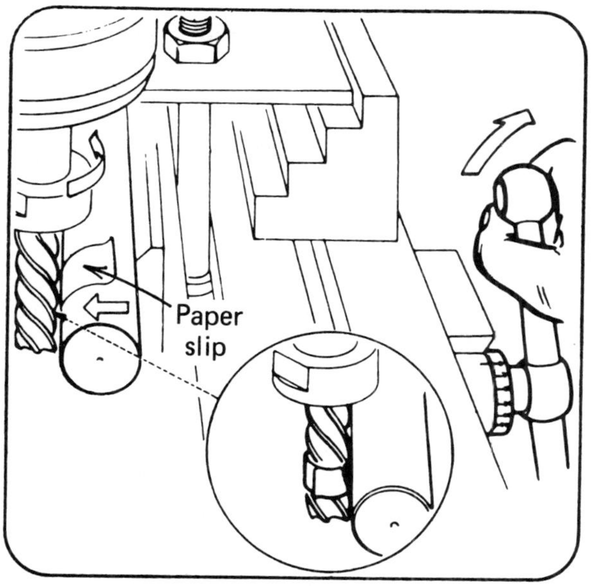

3 Crank table to touch cutter to paper
(a) Be careful that cutter removes paper only without cutting shaft. Stop cutter and remove paper.
(b) Zero table collar.

4 Position cutter to centerline of shaft
(a) Clear cutter over top of shaft.
(b) Calculate amount to move table.

$$\frac{\text{diameter of cutter} + \text{diameter of shaft}}{2}$$

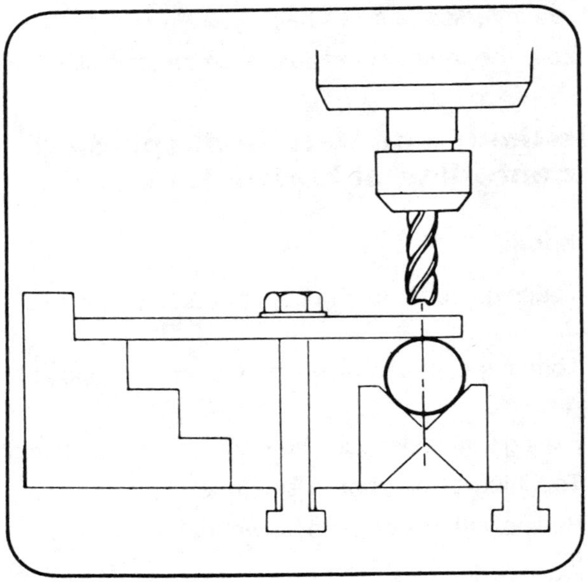

(c) Move table required distance.
(d) Visually check that cutter is centered.

Method 6 Using dial indicator in machine spindle

1 Clamp shaft in position
(a) Select appropriate workholding method.
(b) Correct alignment of shaft is important.

2 Mount dial indicator in machine spindle
(a) Attach magnetic base on spindle nose or use arbor-mounted dial indicator.
(b) Dial indicator stylus toward center of machine spindle.
(c) Disengage machine spindle to allow free rotation by hand.
(d) Visually set machine spindle centered to shaft diameter.

3 Position dial indicator on shaft diameter
(a) Raise table to adjust dial indicator stylus to approximate centerline of shaft.
(b) Adjust dial indicator to engage stylus approximately .020 in. (.5 mm) on shaft.
(c) Raise and lower table until dial indicator stylus is set on centerline of shaft (the point where the highest reading is obtained).
(d) Rotate machine spindle slightly back and forth to find lowest reading on dial indicator.
(e) Set dial indicator on lowest reading and zero bezel.

4 Check setting at opposite side of shaft
(a) Lower table to clear dial indicator over top of shaft.
(b) Rotate machine spindle 180°.

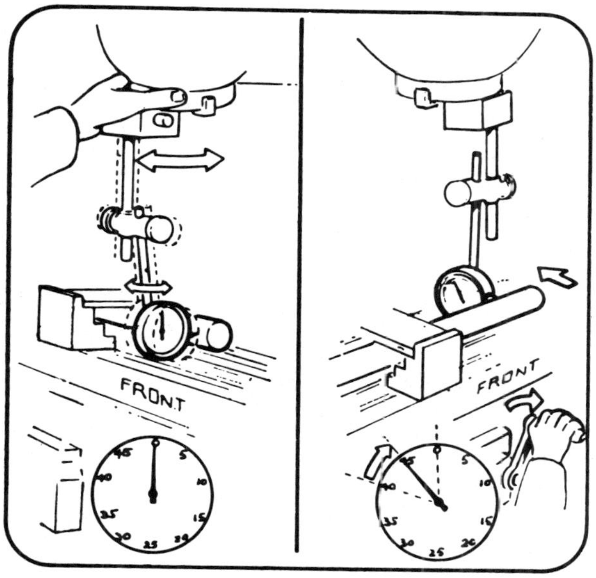

(c) Raise and lower table to position dial indicator at highest reading found on shaft.

(d) Rotate spindle, as before, to find lowest reading on dial indicator.

(e) Note difference in reading from previous side.

5 Adjust table setting

Divide difference of dial indicator readings in half and move table toward lowest reading.

6 Check setting and adjust

Review questions

1 Describe and demonstrate proper table clamping procedure.

2 Describe and demonstrate clamping in vises.

3 What precautions are necessary for clamping thin or hollow workpieces in vises or directly to the machine table?

4 Describe and demonstrate clamping to angle plates.

5 What are stops and how are they used?

6 Describe procedures for edgefinding.

7 Name several tools for edgefinding and demonstrate their uses.

8 Describe and demonstrate the procedure for locating the spindle centerline over a round workpiece.

UNIT EIGHT

Feeds and Speeds in Milling

Cutting speeds and cutter RPM calculations

Correct cutting speeds are critical to cutter preservation and productivity in machining. Cutting speeds refer to the speed at which a point on the milling cutter tooth passes the workpiece in a given period of time. Cutting speeds are measured in surface feet per minute (SFPM). Depending on the cutting tool material, optimum SFPM rates can be determined from the following table.

Cutting Speeds for Milling

	Cutting Speed (SFPM)	
Material	**High Speed Steel Cutter**	**Carbide Cutter**
Free machining steel	100–150	400–600
Low carbon steel	60–90	300–550
Medium carbon steel	50–80	225–400
High carbon steel	40–70	150–250
Medium alloy steel	40–70	150–350
Stainless steel	30–80	100–300
Gray cast iron	50–80	250–350
Bronze	65–130	200–400
Aluminum	300–800	1000–2000

Calculating cutter RPM

Cutter RPM can be calculated by the following formula:

$$\text{RPM} = \frac{4CS}{D}$$

where CS = cutting speed of material being machined
D = diameter of milling cutter

Example: End milling aluminum with a .750 in. diameter end mill.

$$\begin{aligned}\text{RPM} &= \frac{4CS}{D}\\ &= \frac{4(300\ \text{SFPM})}{D}\ \text{(Cutting speed table data)}\\ &= \frac{1200}{.750}\\ &= 1600\end{aligned}$$

Calculating feed rate

Feed rate in milling is measured in inches per minute (IPM) and can be calculated by the formula:

$$\text{IPM} = (\text{FPT})(N)(\text{RPM})$$

where FPT = optimum feed per tooth of cutter (chip load)
N = number of teeth on cutter used
RPM = revolutions per minute of cutter

Example: A four-flute .5 in. diameter high carbon steel shell end mill machining aluminum at .007 in. feed per tooth.

$$\begin{aligned}\text{RPM} &= \frac{4\text{CS}}{D} = \frac{4(300)}{.5}\\ &= \frac{1200}{.5} = 2400\\ \text{IPM} &= (.007\ \text{in. per tooth})\ (4\ \text{teeth})\\ &\quad (2400\ \text{RPM}) = 67\ \text{IPM feed rate}\end{aligned}$$

Recommended FPT rates for materials and cutter types

Optimum feed rates can be determined from the following table.

Material	Cutter Type					
	Face	Helical	Slotting/Side	End	Form	Saw
Aluminum	.022	.018	.013	.011	.007	.005
Brass/bronze	.014	.011	.008	.007	.004	.003
Cast iron	.013	.010	.007	.007	.004	.003
Carbon steel	.012	.010	.007	.006	.004	.003
Tool steel	.010	.008	.006	.005	.003	.003
Stainless steel	.006	.005	.004	.003	.002	.002

Review questions

1 Why are cutting speeds important in machining?

2 How are cutting speeds related to cutter RPM in milling?

3 How are milling cutter RPMs calculated?

4 How are mill feed rates calculated?

5 Calculate the correct RPM and feed rate for milling in cast iron with a form milling cutter.

UNIT NINE

Milling Machine Operations

Climb and conventional milling

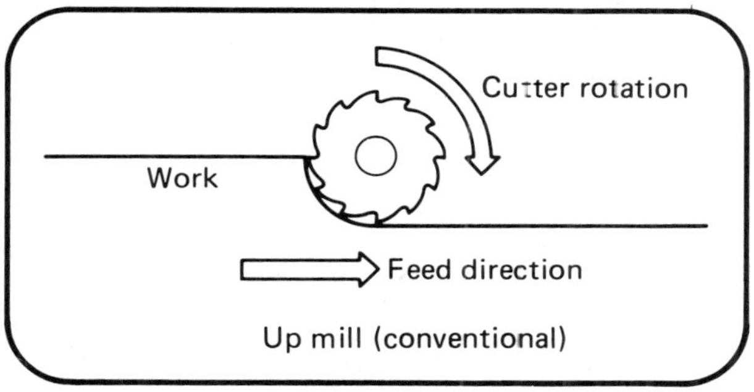

In conventional or up milling, the cutter rotation opposes the direction of feed.

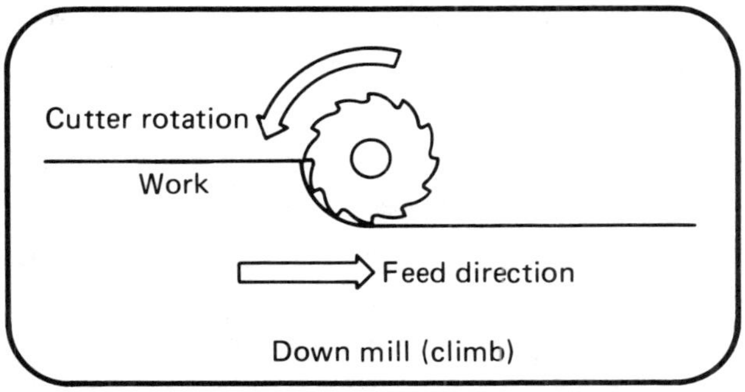

In climb or down milling, the cutter rotation is the same as the direction of feed. The cutter may climb onto the workpiece because of the clearance and resulting backlash of the table or saddle screws and their nuts. Thus can result in a damaged cutter, a bent arbor, or damage to the workpiece.

Climb milling should be avoided on conventional milling machines without backlash compensation. However, very small cuts of depths in the .002–.005 in. range climb milled on the workpiece will often leave improved surface finish. Mills or N/C mills equipped with ball screws will climb mill with no danger because thread clearance and resulting backlash are eliminated.

Milling a block square

1 Set workpiece in vise

(a) Select a pair of parallels high enough to finish

block without cutter hitting vise jaws, but to hold maximum amount of block.
(b) Deburr block and clean vise jaws.
(c) Place smoothest side of block to fixed vise jaw.
(d) Rest block on parallels, approximately centered on length of vise.
(e) Insert a round bar of suitable length, and approximately ¾ in. (20 mm) diameter, between movable jaw and block, and approximately centered on length of block in vise.
(f) Tighten vise and tap workpiece, to bring in same plane as parallels.
(g) Fully tighten vise and tap again.

Note:
The block may not seat tight on the parallels in its rough state.

2 Machine first face
(a) Place a parallel on top of vise body to direct coolant back to table.
(b) Remove enough material to give a good clean face using plenty of coolant.
(c) Clear chips from vise.
(d) Check surface finish.

3 Reset block in vise
(a) Remove block from vise.
(b) Clean vise and parallels, and remove burrs from block.
(c) Place machined side of block to fixed vise jaw, insert round bar, and tighten as before.
(d) Check machined side of block with .0015 in. (.04 mm) feeler gage to ensure that it is square to fixed vise jaw.

Safety

Be careful of your hands when handling the block. Ensure that all guards are in position before starting the machine.

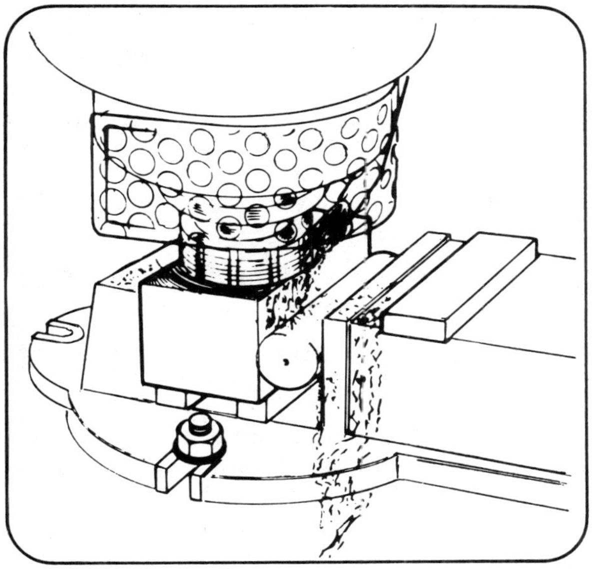

4 Machine second face

5 Check block

(a) Clear chips and remove block from vise.
(b) Deburr block and wipe clean machined faces.
(c) Check the two machined faces with a square.

6 Machine third face

(a) Replace block in vise, one machined face to fixed vise jaw, other machined face on parallels.
(b) Insert bar and fully tighten vise.
(c) Tap block until it seats tightly on both parallels.
(d) Remove material to bring block to size.

7 Machine fourth face

(a) Clear chips, remove block, and deburr.
(b) Clean vise and parallels.
(c) Replace block with remaining unmachined side to cutter.
(d) Tighten vise and seat block tight on parallels.
(e) Machine to finished size.

Note:
A bar is *not* used when machining the last side. When machining material such as phosphor bronze the block should be rough machined on all four sides, leaving approximately .020 in. (.5 mm) all around for finishing.

General Note:
Although the illustrations show machining on a vertical milling machine, the same method is employed on a horizontal milling machine.

Protection of workpieces

1 Remove workpieces carefully and place in containers to prevent damage during transportation.

2 Place flat workpieces either on edge or stacked with machined faces in contact.

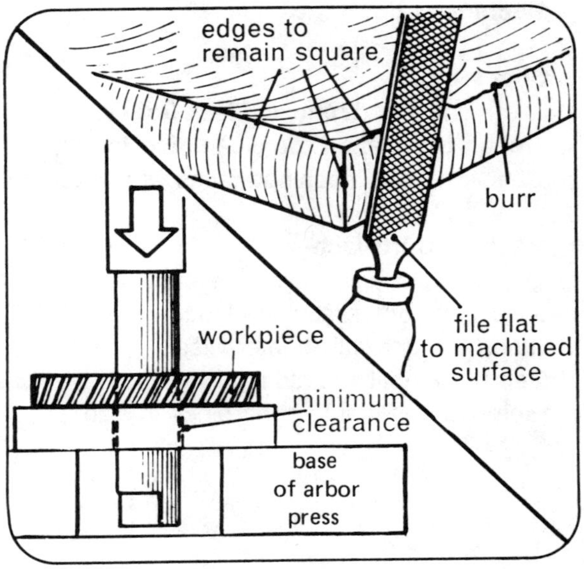

3 Confine deburring operations to removal of burrs. Ensure that corners demanding square edges remain square.

4 Empty pockets or bores of chips. Brush away dust.

5 Drain coolant away and wipe surfaces. Where required, smear with oil.

6 Use an arbor press with full support to the workpiece to prevent distortion.

> **Safety**
>
> Beware of sharp edges.

Surface finish

1 Accurate measurement is controlled by the quality of the surface finish.

2 On a surface finish of desired quality, tooling marks are barely discernible to the naked eye and have characteristics as below.

3 Check the surface finish with comparison plates visually and by drawing a fingernail across the cut direction.

Note:
In face or end milling, a difference will be seen and felt in two directions of the surface. The width between lines of finish is determined by cutter pitch, speed, and feed.

(a) Face or end milling

(b) Side milling of face or end mill

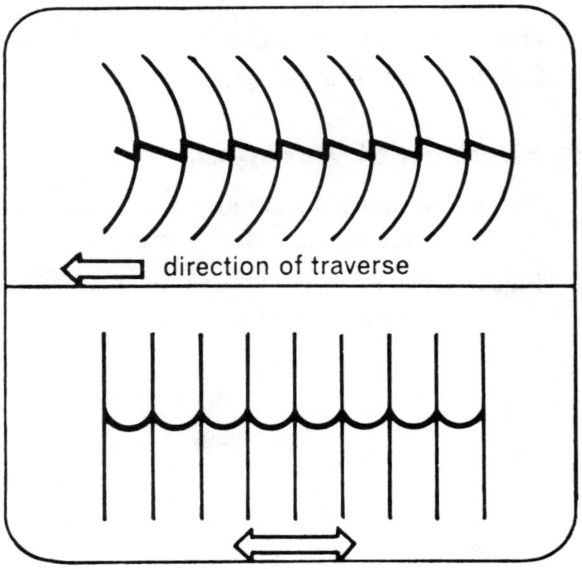

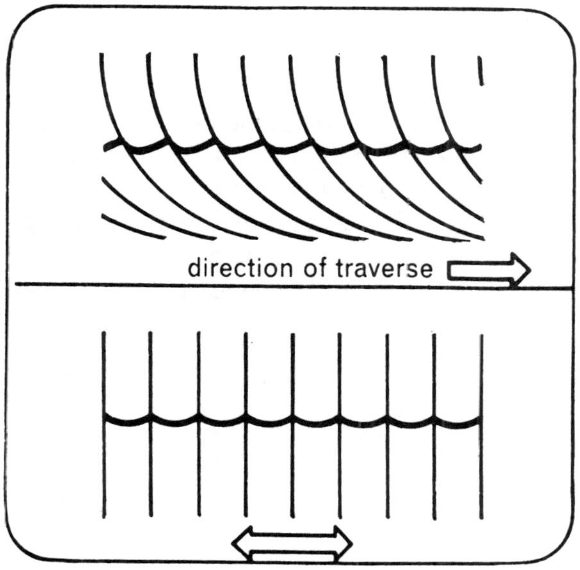

(c) Side milling of side and face cutter

(d) Peripheral milling of side and face cutter

Checking surface finish using comparison plates

Comparison plates are a range of metal blocks, machined to various surface finishes, with the finish of each one marked in microinches or micrometers.

1 Visual check

(a) Select appropriate comparison plate for surface finish required.
(b) Visually check workpiece and plate alternately.
(c) Check from as many different angles as possible, holding plate near workpiece.
(d) Note any cutter marks on workpiece, and if they are deeper or more irregular when compared with plate, surface finish is not up to standard.
(e) Compare fineness or sheen of workpiece with plate. If it is coarse or dull, it is not to standard.

Note:
The surface finish of the workpiece should be at least as good as, preferably better than, the appropriate comparison plate.

2 Check by feel

(a) Draw tip of finger lightly over comparison plate.
(b) Repeat operation over face of workpiece, and note any coarseness or excessive rubbing.
(c) Check at various angles across plate and workpiece alternately. Workpiece should feel at least as smooth as plate and preferably smoother.

Milling a shoulder

1 Mount workpiece

(a) Select suitable workholding method, so that finished depth of cut can be obtained, without hitting workholding equipment or machine.

(b) Ensure that direction of cut is toward a fixed part of workholding and preferably to rear of machine. Example: Toward stops, clamps, or fixed vise jaw.

2 Select cutter

Select cutter (face or end mill) of sufficient diameter to cover full face of shoulder to be machined.

3 Position cutter to workpiece

(a) Start machine, lightly touch periphery of cutter, to reference face of workpiece, and zero micrometer collar.

(b) Move table required amount, less .010 in. (.25 mm), to side face of shoulder. Lock slide.

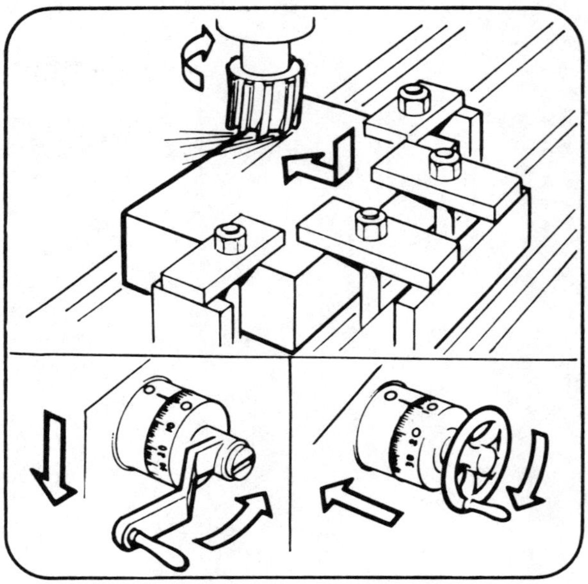

(c) Lightly touch surface of workpiece with face of cutter, and zero collar.

4 Cut out shoulder

(a) Adjust table to remove material in even cuts, using plenty of coolant.

(b) Check side of face of shoulder after a depth of approximately ⅛ in. (3 mm) has been reached.

(c) Check depth of shoulder, when within .020 in. (.5 mm) of finished size.
(d) Adjust collar for both side face and depth and finish machine.

Note:
If an end mill is employed, the full depth less .010 in. (.25 mm) is first set, and the material removed by the side of the cutter. The operation may be performed on a horizontal milling machine by setting up in the same manner used when an end mill or side and face cutter is employed.

Milling parallel faces centered on round stock

1 Mount workpiece
(a) Select suitable parallel, with thickness less than diameter of workpiece.
(b) Check height of parallel to ensure that centerline of workpiece is below top of vise jaws and is set as low as possible to machine face without cutter hitting jaws.
(c) Clean and deburr workpiece, and mount in vise with parallel centered to diameter.
(d) Position parallel so that workpiece extends over one end, for ease of measurement.
(e) Tighten vise, and tap workpiece with soft hammer to seat tightly on parallel.

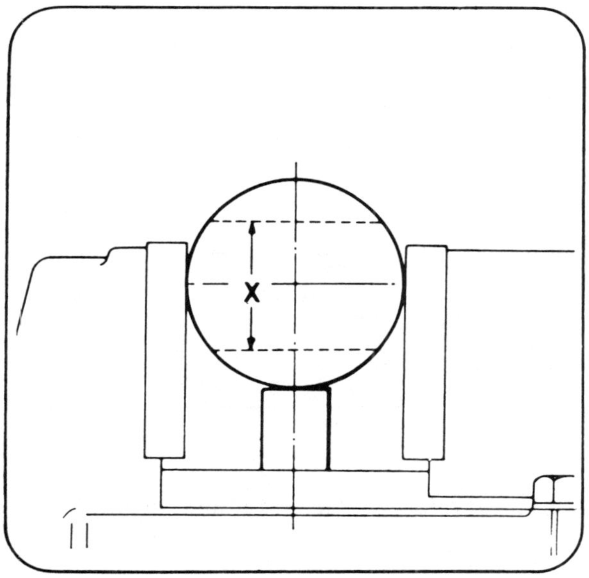

2 Select cutter
Select a suitable face mill with diameter sufficient to cover face of workpiece at finished size.

3 Position cutter to workpiece
(a) Adjust cross slide to position cutter approximately centered to workpiece. Lock slide.
(b) Move longitudinal slide to position front of cutter over end of workpiece.

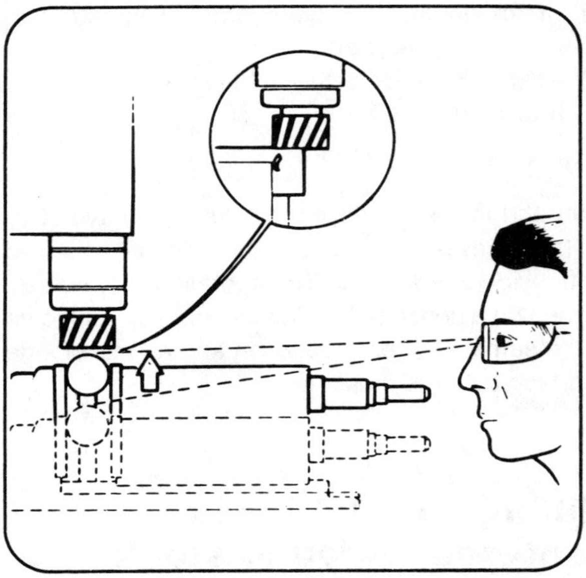

(c) Raise table to bring face of cutter within approximately ¼ in. (6 mm) of workpiece.
(d) Start machine and adjust vertical slide gently to touch cutter to workpiece, marking workpiece lightly.
(e) Set vertical index at zero.

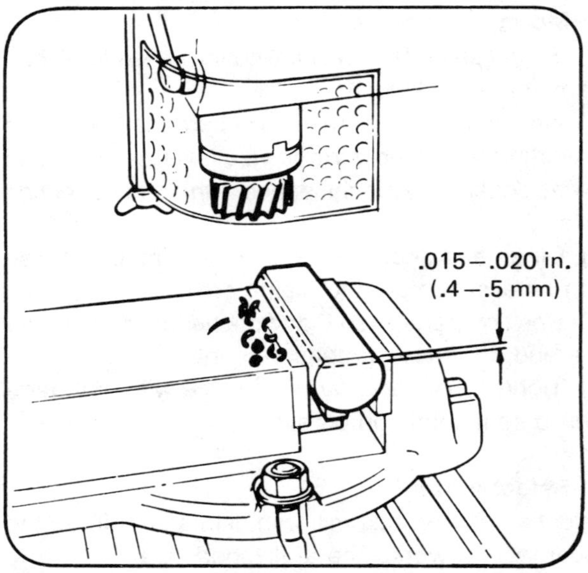

4 Rough machine first face
(a) Calculate amount of material to be removed.

Example:

$$\frac{\text{bar diameter} - \text{finished size}}{2}$$

(b) Rough machine face to within .015 in.(.4mm).015–.020 in.(.4–.5 mm) of finished size, using plenty of coolant.

Safety

Ensure that all guards are in position before starting the machine.

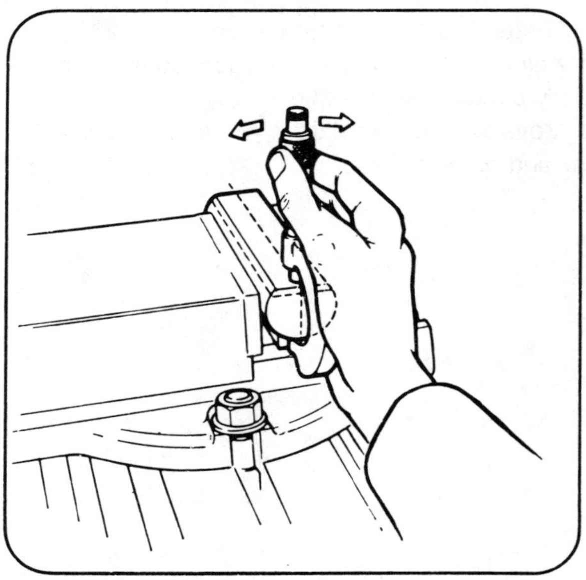

5 Position micrometer and check size
(a) Move longitudinal slide to move cutter well clear of workpiece.
(b) Clean chips and coolant from workpiece.
(c) Position micrometer centered on radius of workpiece.
(d) Rock micrometer slightly while rotating barrel, to aid in locating spindle flat on top face.
(e) Retain micrometer reading when removing from workpiece.

6 Finish machine first face
(a) Adjust vertical slide and finish first face. (Example: micrometer reading minus finished size.)

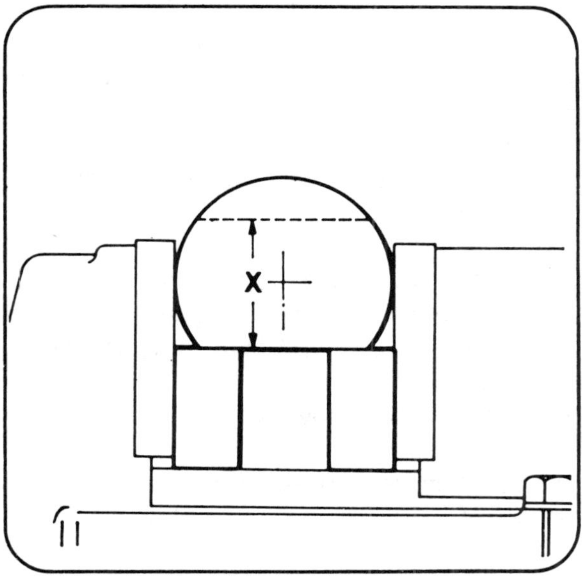

7 Mount workpiece for opposite face

(a) Select suitable pair of parallels and mount workpiece on finished face.

(b) Tighten vise, and tap workpiece with soft hammer to seat tightly on parallels.

8 Machine second face

(a) Repeat steps 4, 5, and 6.

(b) Remove workpiece and deburr.

Safety

Be careful of your hands when deburring the workpiece. Wear safety glasses when cutting.

Note:

When machining materials that tend to distort on cutting, such as phosphor bronze and low grade, low carbon steel, or workpieces of thin finished size, the workpiece should first be rough machined on each side. The finished size then is obtained by one or more light cuts, taken alternately over each face.

The operation may be carried out on a horizontal machine, in the same manner. A slab or roller mill is employed in place of a face mill.

Milling an angle by worksetting

1 Mark out workpiece

Apply layout dye to workpiece, and scribe required angle in appropriate position.

2 Set workpiece at required angle

(a) Mount workpiece in vise, visually align scribed line with top of vise jaws.

(b) Set workpiece with line approximately ⅛ in. (3 mm) above vise jaws.

(c) Hand tighten vise, to hold workpiece steady.

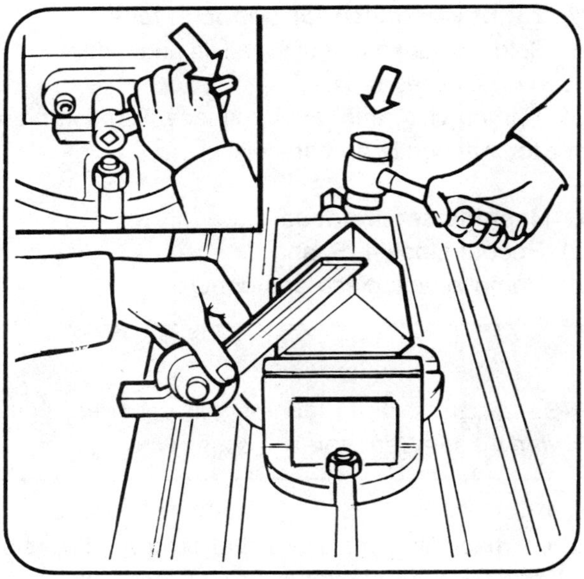

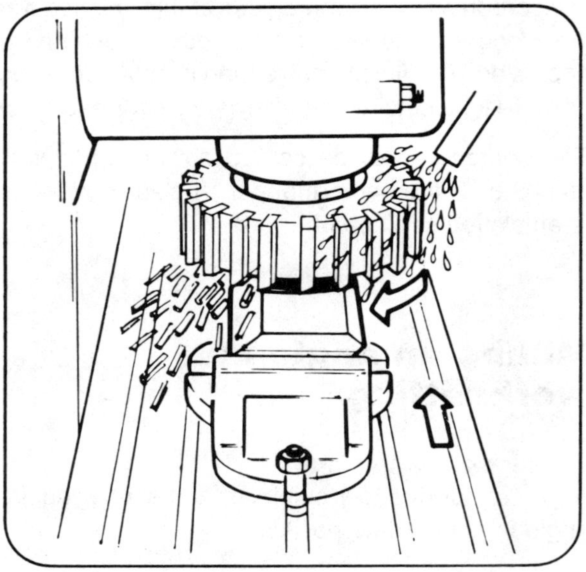

(d) Set protractor at required angle, hold base firmly on machine table, and check face alignment.
(e) Tap workpiece with soft hammer until exact alignment is obtained.
(f) Fully tighten vise.
(g) Recheck with protractor, to ensure that workpiece has not moved.

3 Position cutter to workpiece
(a) Raise table to bring bottom face of cutter within approximately 1/16 in. (1.5 mm) of highest point of workpiece.
(b) Adjust saddle to cover whole face of angle with cutter, and lock slide.
(c) Start machine, and remove material until scribed line is split.

Note:
For the first few cuts, offset the workpiece to the cutter centerline to ensure that the direction of thrust in cutting is toward the fixed vise jaw. This tends to prevent the workpiece from tipping while cutting is in progress.

After the first cut, the surface area will become greater, and the depth of cut should be reduced for subsequent cuts.

Angles may be machined on a horizontal machine by setting the workpiece at the required angle and removing material with a slab mill or side and face cutter. Small angles may be machined with an angle cutter.

Milling a slot

1 Set vise parallel

2 Set for position of slot with setting bar
(a) Open vise jaws to give plenty of clearance for setting bar.
(b) Engage setting bar and center approximately on vise.
(c) Check setting bar for concentricity.
(d) Raise table to engage setting bar to register on *fixed* vise jaw.

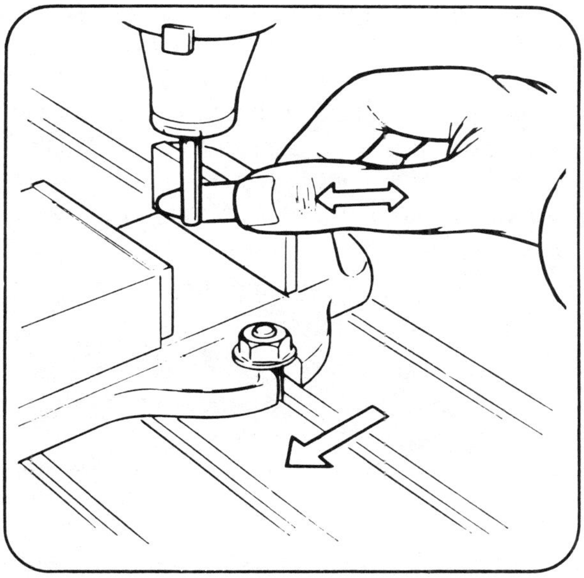

(e) Move table across until bar is within 1/16 in. (1.5 mm) of the fixed vise jaw.

(f) Insert a .001 in. (.025 mm) feeler gage, and carefully move table across to trap feeler between bar and fixed vise jaw. Rock feeler gently until tightness is obtained.

(g) Zero collar.

(h) Move table back required distance *MINUS* half the diameter of setting bar, to give correct center position for slot.

(i) Move table at least .030 in. (.8 mm) past required setting and then move it back to correct setting, to take up any free movement (backlash) in table screw.

Note:
The slot may be centered to the workpiece or a specific distance from one side.

3 Set workpiece in vise

(a) Ensure that workpiece is free from burrs and vise is clean.

(b) Seat on pair of parallels (if necessary) to bring top of workpiece approximately 1/4 in. (6 mm) above vise jaws. Parallels should be tight after seating.

4 Set for depth and cutting slot

(a) Engage and check concentricity of cutter.

(b) Touch top of workpiece (with cutter revolving), marking workpiece very lightly.

(c) Set vertical collar at zero.

(d) Set traverse stops at each end of table, to stop traverse when cutter has completely cleared workpiece at each end.

(e) Remove material in even cuts. Take cut alternately from ends of workpiece.

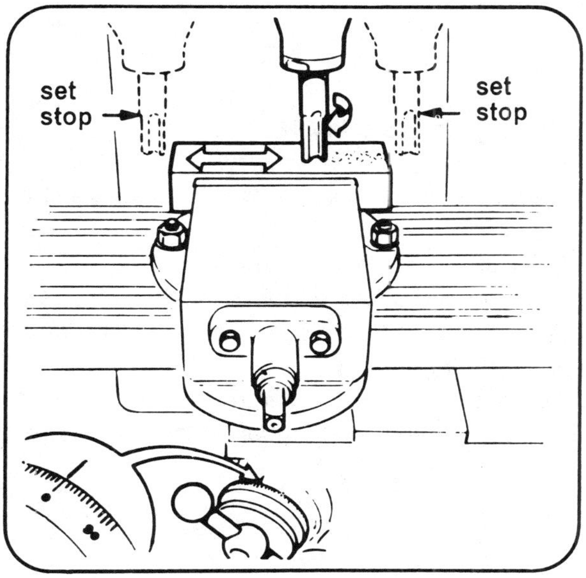

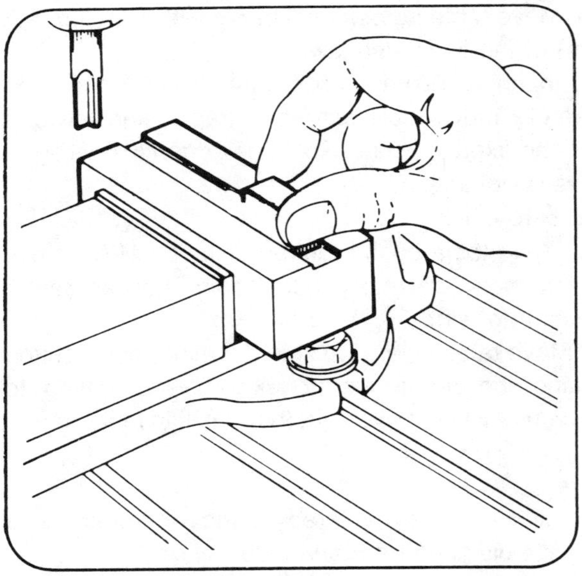

5 Check width, depth, and position of slot
(a) Check width of slot with gage blocks, after a depth of 1/16 in. (1.5 mm) has been obtained.
(b) Check position of slot with a micrometer or vernier after a depth of 1/8 in. (3 mm) has been obtained.
(c) Check depth of slot with depth micrometer or vernier when within .010 in. (.25 mm) of full depth on collar.

Safety

The cutter must be well clear of the workpiece and stopped and any chips removed before the slot is checked.

Note:
It is possible to cut many types of slot on a vertical mill. The slot may not cover the whole workpiece, in which case an index reading must be taken or a stop set for the blind end (see "Milling a keyway").

Also the stub arbor may be employed together with a side and face cutter. Round workpieces are usually set in the dividing head for this operation.

The operation may be carried out on a horizontal machine in the same manner, when a side and face or slot cutter is employed. Alternatively the workpiece may be clamped to the rear of the machine table, or to an angle plate, and a cutter mounted in the machine spindle.

Milling a keyway

1 Repeat steps 1, 2, and 3 as for "Milling a slot"

2 Set cutter to length of keyway
(a) Engage and check concentricity of cutter.
(b) Touch end of workpiece with cutter and paper (as in centering cutter).
(c) Move table required length from end of workpiece, *PLUS* diameter of cutter, using table index.

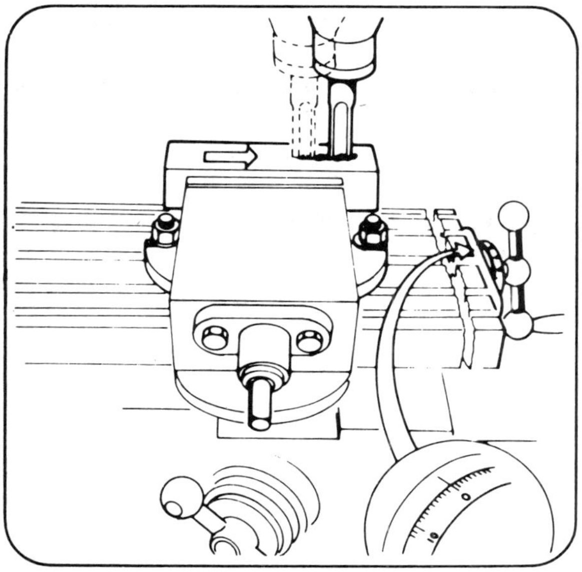

(d) Set collar at zero.

(e) Move table to required length of keyway *MINUS* diameter of cutter, and mark collar.

Note:

Backlash must be accounted for when moving in the reverse direction.

3 Set for depth and cut keyway

(a) Touch top of workpiece with revolving cutter, marking workpiece very lightly.

(b) Set vertical collar at zero.

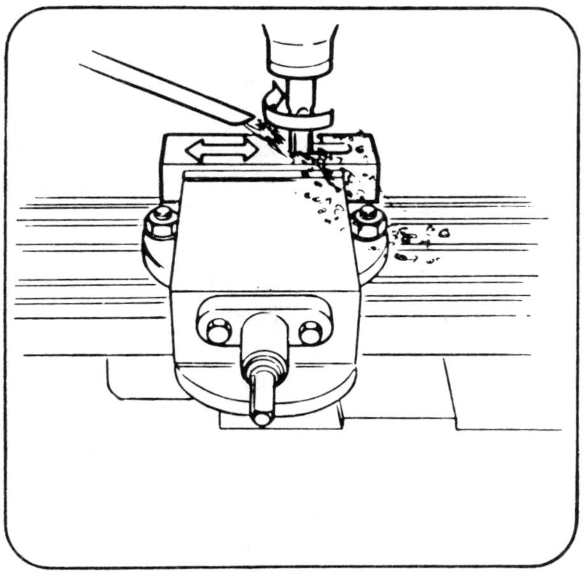

(c) Remove metal in even cuts until required depth is reached.

(d) Use plenty of coolant to aid in chip clearance.

Note:

The operation can be carried out on a horizontal machine by selecting a workholding method to position the workpiece at the rear of machine table (e.g., dividing head, vise, clamps, and the slot drill mounted in the machine spindle).

Alternatively:

If the keyway runs through the workpiece endwise, a side and face or slot cutter is mounted on the arbor, and the workpiece is approximately centered across the machine table.

Checking keyway dimensions

1 Check width of keyways

(a) Select appropriate gage blocks to build up to required size.

(b) Clean gage blocks on clean chamois or soft cloth.

(c) Build up stack by wringing together with slight rotary motion. When correctly built up, blocks should adhere to each other.

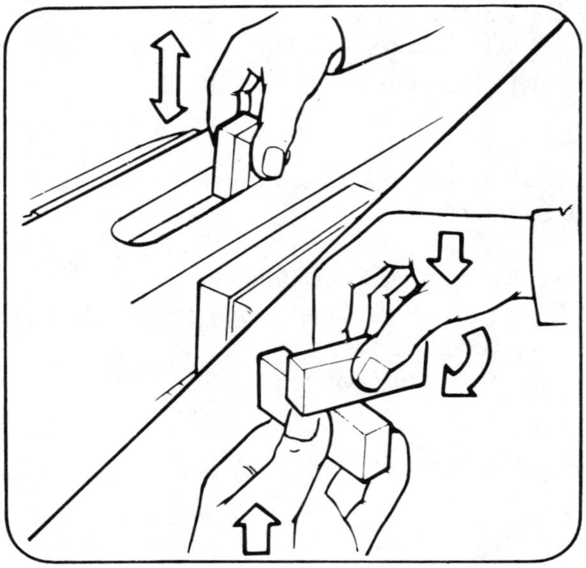

(d) Place gages in keyway, ensuring that keyway is clean and free from burrs.
(e) Blocks should be a slightly tight slide fit, for whole depth of keyway.

Note:
An adjustable parallel may also be used to gage keyway width.

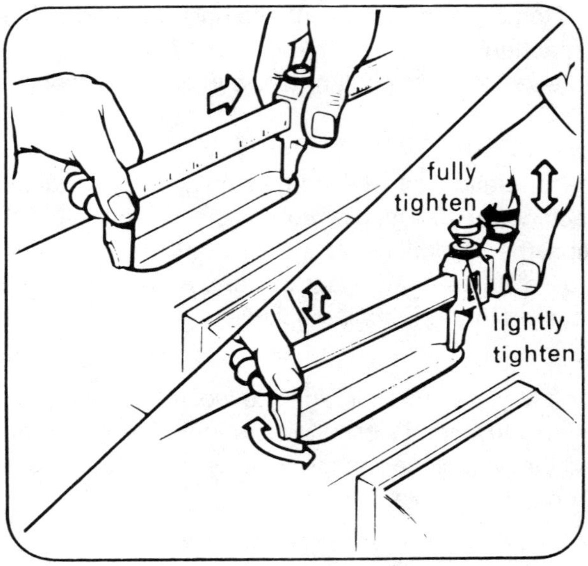

2 Check length of keyway

(a) Select a caliper of appropriate size.
(b) Hold fixed jaw in left hand and sliding jaw in right hand.
(c) Place caliper point of fixed jaw to one end of keyway and slide other jaw to touch opposite point to opposite end of keyway.
(d) Tighten locking screw directly above fine adjustment screw and apply light pressure to locking screw on movable arm (sufficient to allow freedom of movement for fine adjustment).
(e) Turn fine adjustment screw, and with slight up and down movement, obtain a smooth slide fit. Care should be exercised not to strain caliper by overtightening.
(f) Apply slight rotary movement to caliper, to ensure that reading is taken at maximum width of keyway.
(g) Lock sliding jaw, remove caliper from workpiece, and note reading.

Note:
Keyways are normally checked for width and length, after a depth of approximately 3/32 in. (2 mm) has been machined, and checked again after the operation has been completed.

3 Check depth of keyway

A Keyways in shafts, etc.

(a) Select a parallel pin or roller having a diameter less than width of keyway and a micrometer of appropriate range.

(b) Place roller lengthways, on bottom of keyway.
(c) Open micrometer until gap between spindle and anvil is sufficient to pass over workpiece and roller.

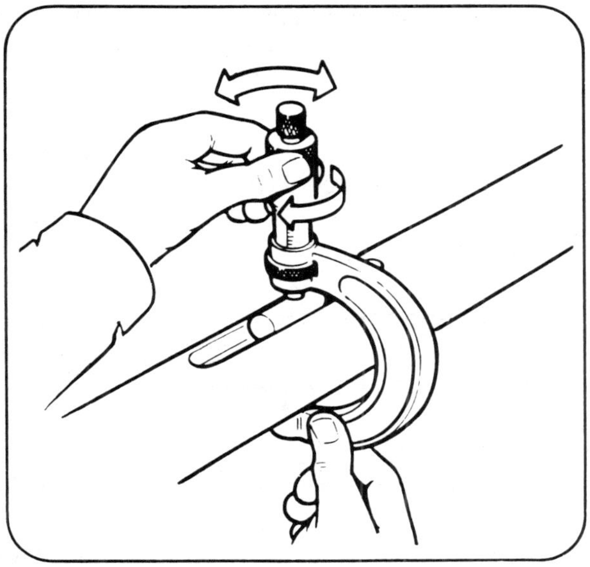

(d) Position micrometer with thimble in vertical plane and gently turn until spindle and anvil just touch workpiece and roller. Feel contact on both sides and move micrometer back and forth a little over bottom of diameter of shaft and top of diameter of roller.
(e) Adjust to final setting by movement of ratchet, and tighten locking ring.
(f) Remove from workpiece and note reading. Depth of keyway equals micrometer reading minus diameter of roller subtracted from full diameter of workpiece.

B Rectangular workpieces

(a) Select suitable micrometer and place anvil under bottom of workpiece.

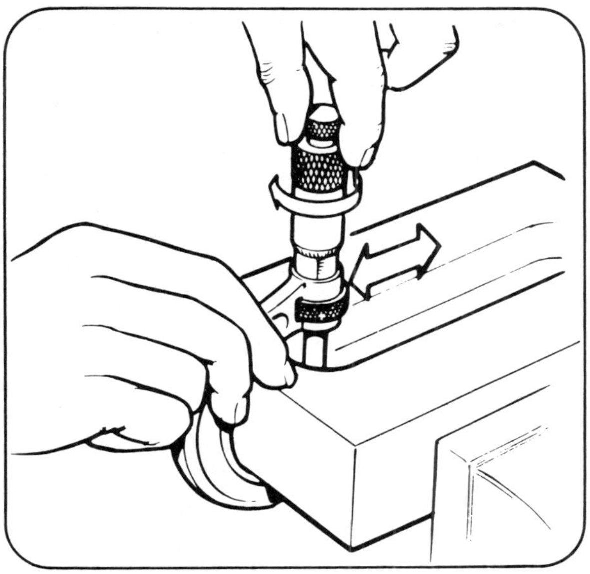

(b) Slide micrometer gently to obtain good seat, and take final setting by movement of ratchet.
(c) Tighten locking ring. Note reading. Depth of keyway equals whole dimension of workpiece minus micrometer reading.

Alternatively:
(a) Select a depth micrometer and assemble with appropriate rod.
(b) Check micrometer zero reading by placing base flat on machine table, and gently turn thimble until end of spindle lightly touches table. Micrometer should read zero at this point. Adjust if necessary.

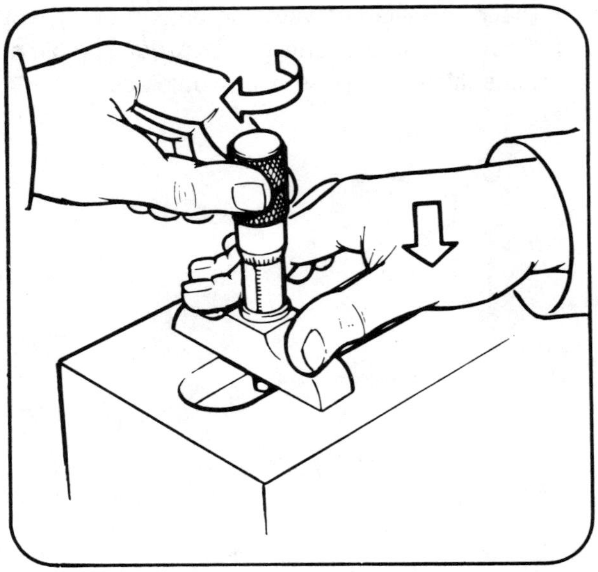

(c) Position base of micrometer flat on top of workpiece, ensuring good seat.
(d) Touch rod very lightly. Excessive pressure will cause base to lift and give false reading.

Note:
If the width of the keyway is such that the spindle of the micrometer will not enter, a roller is employed as in step 3A (shafts, etc.). The micrometer reading is taken from the top of diameter of the roller to the base of the workpiece. The same applies for the checking of slots.

Milling a T-slot

T-slots are cut in two operations; first a slot is milled with an end mill, and then the T-slot cutter is employed to make the cut at the bottom of the first slot.

Note:
The machine will be set in the correct centered position from the first operation.

1 Mill slot

2 Select and set cutter
(a) Select T-slot cutter and check diameter and width.
(b) Engage and check concentricity of cutter.

3 Set for depth
(a) Place piece of wet tissue paper on top of and near end of workpiece.
(b) Move table up (with cutter revolving) until cutter lightly touches paper.
(c) Stop cutter and move clear of workpiece.
(d) Set vertical collar at zero.
(e) Move table up the required distance and reset collar to zero.
(f) Lock vertical slide.

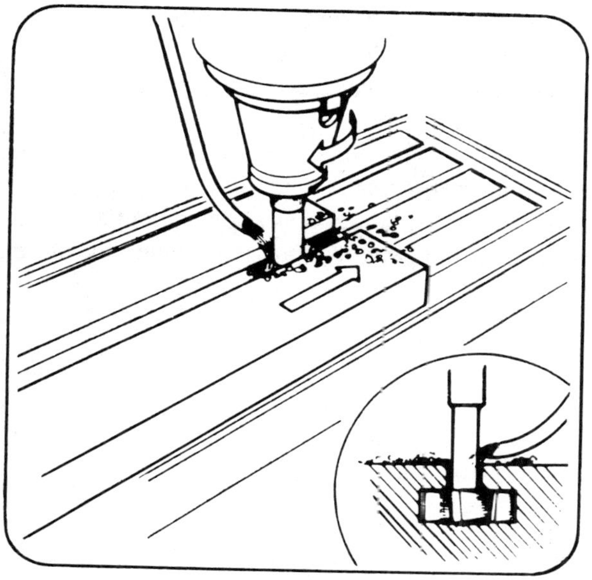

4 Cut T-slot

(a) Select appropriate feed and speed.
(b) Begin cutting and visually check that cutter is giving a good match to bottom of slot in workpiece.
(c) Adjust depth setting if cutter is cutting in, or leaving a step at bottom of slot.
(d) Use plenty of coolant. Also, a brush may be used to aid in chip clearance. It is important to keep chip buildup to a minimum on this operation.

> **Safety**
>
> Keep your hands well clear of the cutter when removing chips with a brush.

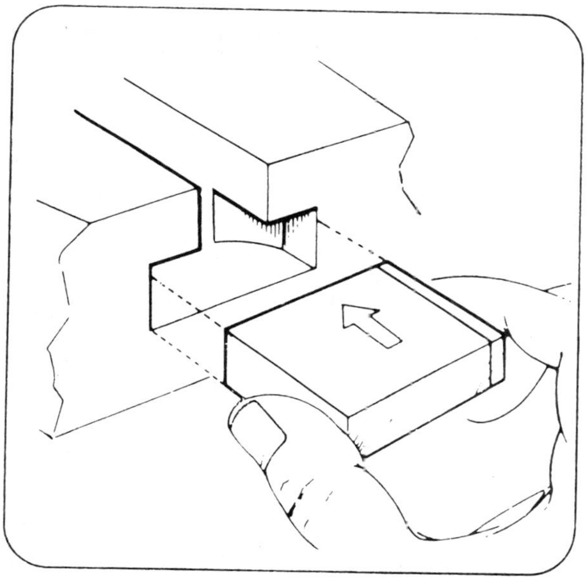

5 Check T-slot

(a) Traverse cutter in until a sufficiently flat side is obtained on sides of slot to admit gage blocks.
(b) Stop cutter and move well clear of workpiece.
(c) Remove any chips or coolant.
(d) Check slot with gage blocks, caliper, or adjustable parallel and micrometer.

Note:
The operation is carried out on a horizontal machine by selecting an alternative method of workholding. For example, workpiece is clamped to the machine table or an angle plate, facing the vertical column, and the T-slot cutter mounted in the machine spindle.

Center drilling

1 Mount workpiece

(a) Select suitable workholding method, ensuring that workpiece is clear of machine table to allow drill to pass through after center drilling.
(b) Align reference face square or parallel.

2 Position machine spindle to workpiece

(a) Mount suitable setting bar in machine spindle.

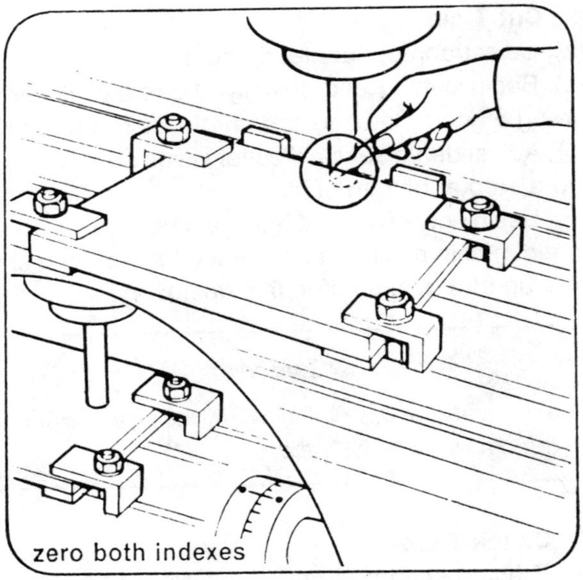

(b) Position spindle center to correct position from reference face for first hole.
(c) Set table and saddle collars at zero and lock slides.

3 Select center drill
Select a suitable center drill, shank diameter as near as possible, but slightly less than, diameter of hole to be drilled.

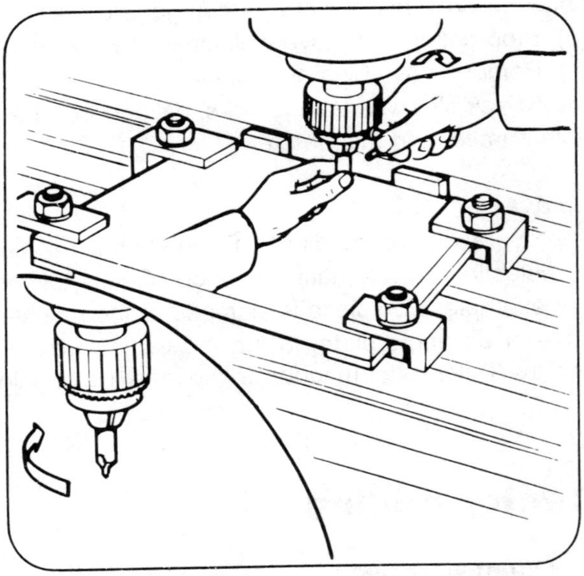

4 Mount center drill
(a) Mount center drill in chuck and fully tighten with appropriate key.
(b) Start machine and visually check that center drill is running true.

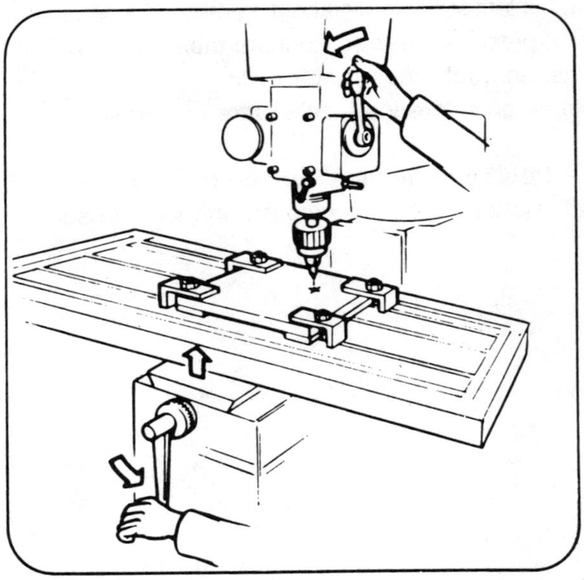

5 Center drill workpiece
(a) Position vertical slide to bring point of center drill within approximately 1 in. (25 mm) of face to be drilled.
(b) Start machine and lower quill to touch point of center drill on workpiece.
(c) Apply small amount of coolant and hand feed quill to sink pilot end of center drill in workpiece.
(d) Raise quill slightly to clear chips; apply more coolant.

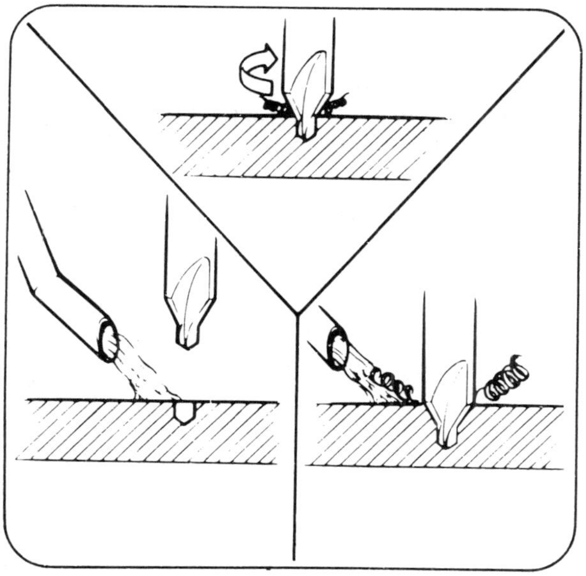

(e) Apply additional pressure to quill and sink center drill into workpiece, to 90% of large angle.

Note:
The power feed is not normally used for center drilling unless a very large center drill is employed. Center drilling is carried out on a horizontal machine in the same manner, except that alternative workholding methods may have to be used, and the saddle is employed to feed the work during the cut.

Drilling

1 Select drill
(a) Select drill of required size and check with micrometer or drill gage.
(b) Visually check drill to ensure that it is ground correctly and that cutting edges are not chipped or damaged.

2 Mount drill
(a) Select suitable sleeve to build up drill taper to size of arbor.
(b) Ensure that tapers on sleeve, drill, and arbor are clean and free from burrs.
(c) Align drill tang with slot in sleeve and push drill into sleeve, ensuring that it is tight.
(d) Push sleeve into arbor, twisting sleeve around until tang enters slot in arbor.
(e) Retain this position, and pull sleeve back approximately 1 in. (25 mm). Then drive home with a sharp thrust.

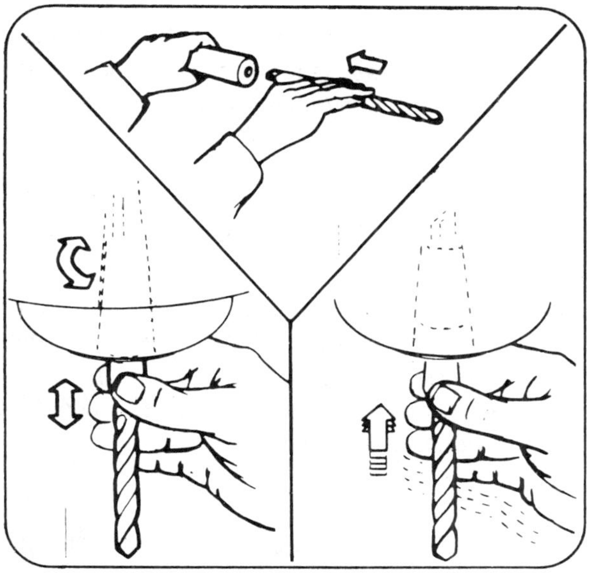

Safety

Always hold drills by the neck, above the flutes, when engaging in sleeves or arbor.

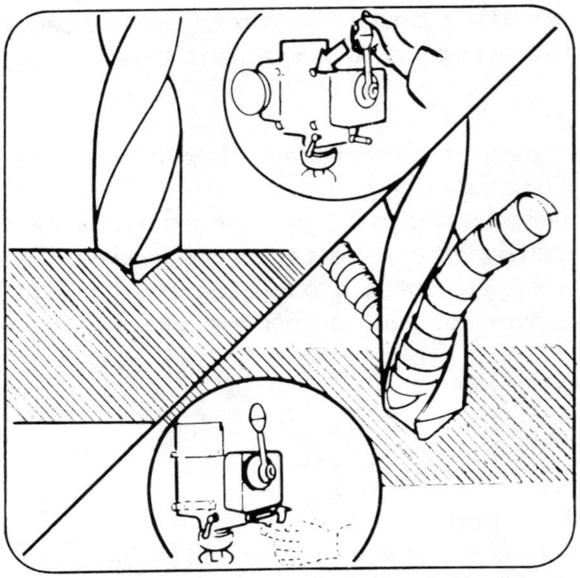

3 Drill through holes

(a) Lower quill to bring point of drill to center drill hole using hand feed.

(b) Engage power feed and apply coolant.

(c) Back out drill at intervals if distance of through holes is great.

(d) Visually check chips. If blue or discolored, reduce feed.

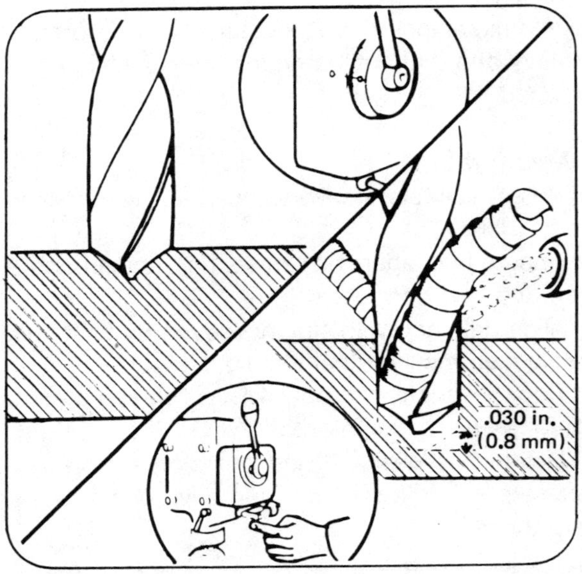

4 Drill blind holes

(a) Lower quill until drill makes full cut on seating. That is, when whole diameter of drill is entering workpiece.

(b) Set quill collar at zero.

(c) Engage power feed and allow drill to enter workpiece to within .020 in. (.5 mm) to .030 in. (.8 mm) of full depth.

(d) Disengage power feed and hand feed to full depth using collar.

Note:
When drilling through holes by hand feed, if resistance on the hand feed is felt to be reducing, reduce the pressure and lightly feed the drill through.

Reaming

1 Select reamer

(a) Select reamer of required size. Check diameter with micrometer.

(b) Visually check flutes and chamfer to ensure that they are not chipped or worn.

2 Mount reamer

(a) Select suitable sleeve to build up reamer taper to size of arbor taper.

(b) Clean tapers on reamer, sleeve, and arbor.

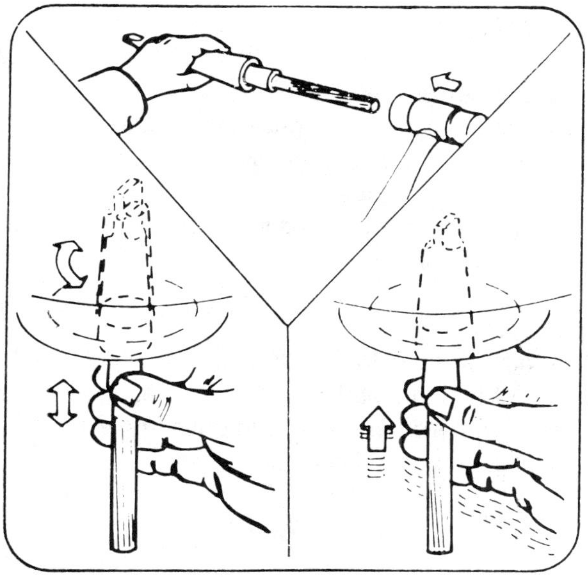

(c) Align tang on reamer with slot in sleeve and seat reamer into sleeve. Tap end of reamer with soft hammer to ensure taper seating.
(d) Mount in arbor.

3 Ream holes
(a) Lower quill to bring end of reamer within approximately .015 in. (.4 mm) of workpiece.
(b) Engage power feed and apply coolant.

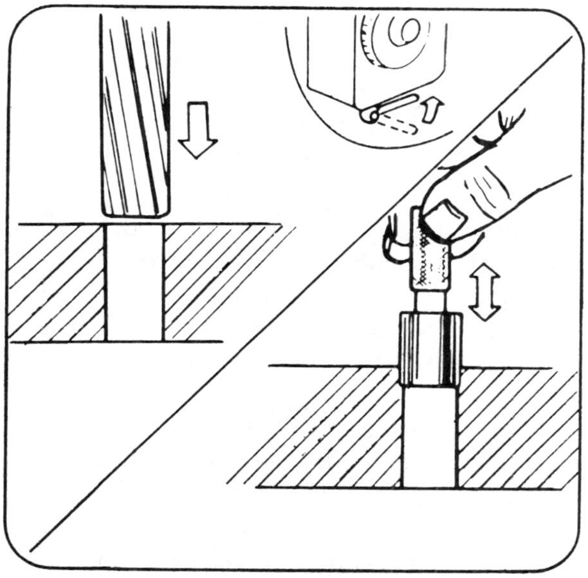

(c) Check diameter of hole with plug gage after depth of approximately ⅛ in. (3 mm) has been reached.
(d) Power feed reamer through workpiece, until chamfer is clear of bottom face. Use fairly heavy feed rate and spindle RPM one-half to two-thirds of drilling RPM.

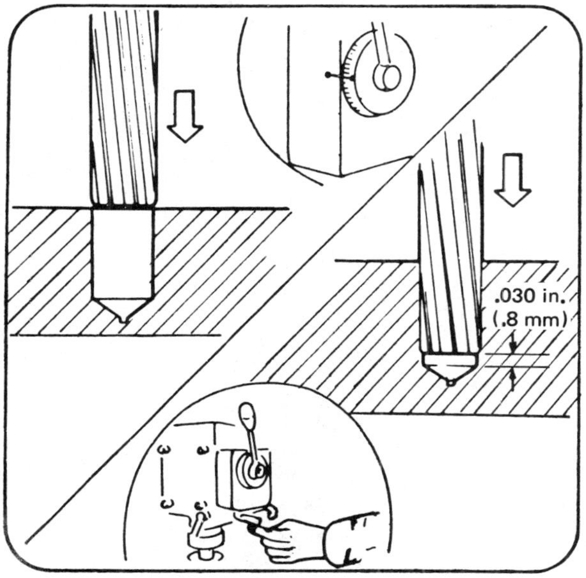

4 Ream blind holes
(a) Lower quill until full diameter of reamer is just touching top of diameter of drilled hole.
(b) Set quill collar at zero.
(c) Power feed reamer to within approximately .030 in. (.8 mm) of full depth, using plenty of coolant.
(d) Hand feed gently, for remaining .030 in. (.8 mm).

Note:
When reaming holes by hand feed, the application is the same. Light, even pressure should be maintained on the feed handle.

Tapping

1 Mount center

(a) Select a suitable spring-loaded center.

(b) Build up taper on center with suitable sleeve(s), to mate with arbor taper.

(c) Mount center in arbor, ensuring that tang on sleeve is engaged in slot in arbor.

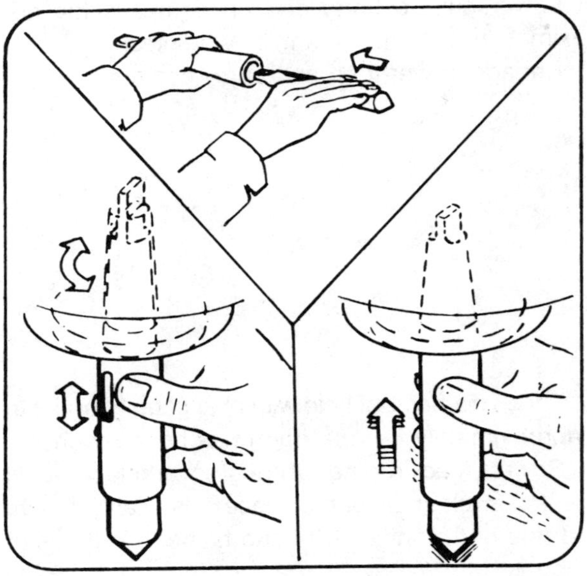

2 Insert tap

(a) Attach wrench to first tap, and apply a small amount of tapping lubricant to end of tap.

(b) Place tapered end of tap in drilled hole and visually align in vertical position.

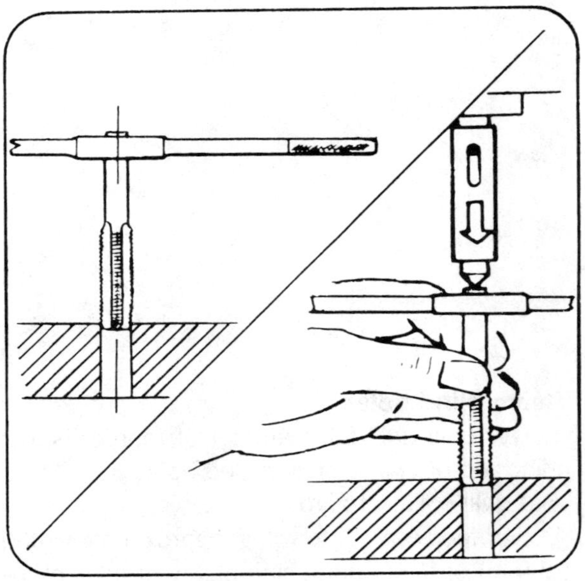

(c) Support tap and wrench with left hand, and drop quill to engage spring center with center on tap.

(d) Apply additional pressure to quill, until spring pushes peg to top of slot in center.

Note:
The machine spindle will be set in the correct centered position from the drilling operation.

3 Tap hole through

(a) Apply light pressure and break chip at regular intervals by turning back approximately a half turn until tap is free.

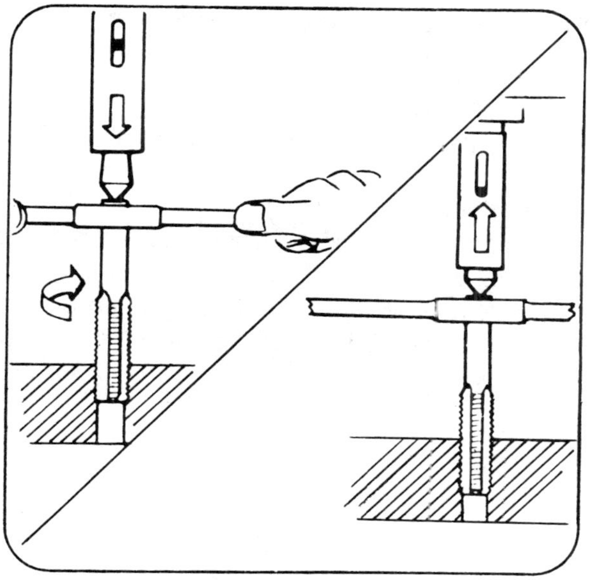

(b) Apply additional pressure to quill when all tension has been taken up in spring center; that is, when peg has reached bottom of slot in center.

Note:
On small diameter holes, the intermediate tap alone may be sufficient to give the required thread. On larger holes however, the starter and plug tap should also be employed.

4 Check threads
(a) Clean tapped hole.
(b) Check size with thread gage.

> **Safety**
> Safety glasses must be worn when using compressed air.

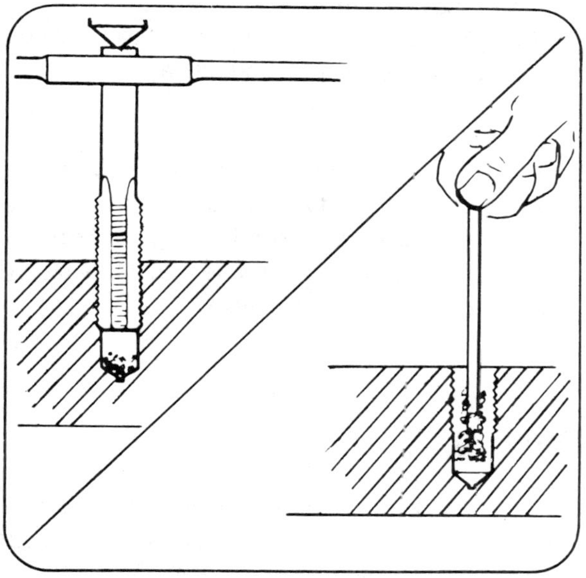

Tapping blind holes

1 Tap hole to depth
(a) Measure length of tap and subtract depth of hole, to obtain remaining tap projection.
(b) Mark tap at required depth.
(c) Proceed as before, keeping hole free from chips by removing tap and clearing chips from hole with magnet or compressed air.
(d) Exercise great care when nearing full depth. When resistance on wrench increases, remove tap and inspect hole.
(e) Use intermediate and plug taps and repeat process for full threads.

Spotfacing

Spotfacing is usually carried out on castings and similar items that require a face for nuts and washers or cap screws. The workpiece position is retained from drilling the hole, and the mill or spotface cutter is employed to carry out the operation. Spotface cutters have a pilot retained with a set screw. The diameter of the pilot must be free fit in the original drilled hole in the workpiece.

Facing with spotface cutter

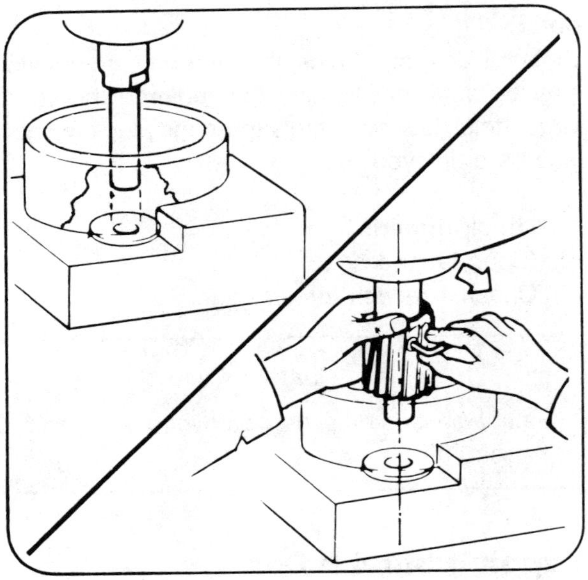

1 Selection of cutter assembly

(a) Select facing cutter and pilot.

(b) Slide pilot on cutter and lock in position.

2 Spotface workpiece

(a) Build up spotface shank with appropriate sleeve(s), and mount in arbor in machine spindle.

(b) Apply a small amount of lubricant to pilot end.

(c) Raise table to bring pilot end within 1 in. (25 mm) of workpiece.

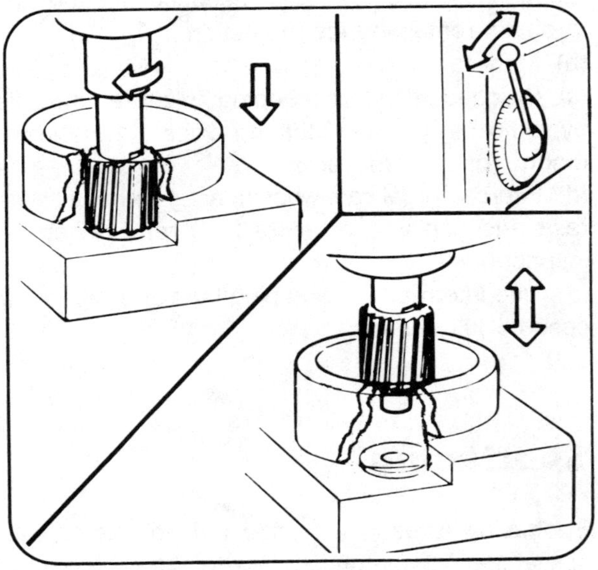

(d) Lower quill (cutter revolving), to enter pilot in hole, and move cutter face to workpiece.

(e) Apply additional pressure to quill, relieving cutter at intervals as it cuts into workpiece.

Note:
On castings and so on with a hard skin or surface, heavy pressure should be applied to allow the cutter to cut under the skin, before forming the face.

(f) Visually check face to ensure that it has cleaned up for whole diameter of cutter.

Boring

Boring holes on the milling machine usually consists of boring cored holes in castings, or holes, counterbores, or recesses in other workpieces that are too large to finish by other methods. Unless a blind bore or counterbore is required, the cutting edge of the tool should be ground to enable the chip to curl away from the finished cut and to reduce the tendency of the tool to spring into the work.

Setting to centerline of bore

1 Position dial indicator

(a) Handcrank to position bore approximately centered to machine spindle.

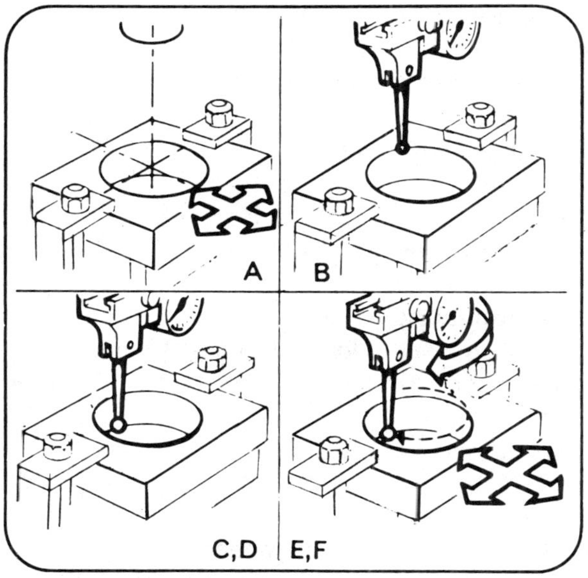

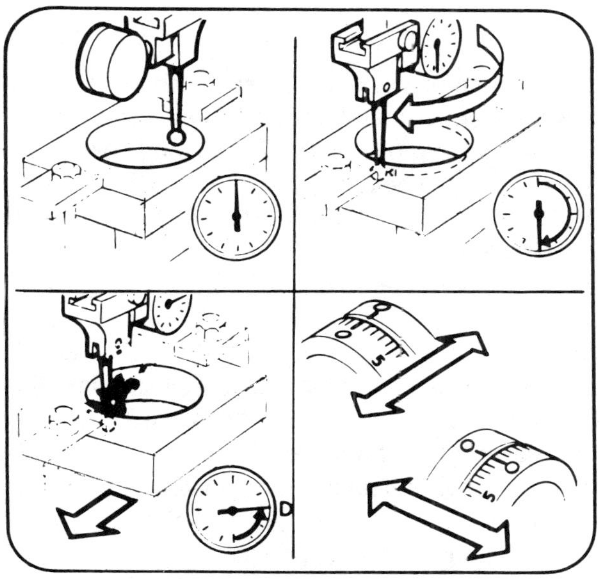

(b) Engage dial indicator base to machine spindle with stylus toward face of bore.
(c) Position dial indicator stylus just above top of bore.
(d) Adjust dial indicator stylus to approximate diameter of bore.
(e) Rotate spindle, by hand or at very slow speed, to see amount of gap between stylus and face of bore.
(f) Move table until gap is approximately even all around bore.

2 Center bore to spindle
(a) Raise table to position stylus approximately ⅛ in. (3 mm) below top of bore.
(b) Adjust dial indicator to engage stylus on wall of bore and zero dial collar.
(c) Rotate spindle. Note reading on dial indicator at opposite sides of bore in line with table. Determine amount of error.

Note:
Concentrate on one axis at a time.
(d) Position dial indicator at lowest point and move table toward stylus a distance of half the noted error.
(e) Repeat checking and adjusting table position until a zero indicator reading is obtained all around bore.
(f) Zero machine collars, noting direction of backlash. The table may now be moved off any distance from zero.

If several workpieces are to be machined with a bore as reference:

1 Select a plug that is a sliding fit in bore.

2 Clamp plug to machine table, ensuring that bolt(s) will not hit workpiece.

3 Align plug and zero collars on centerline.

4 Position workpiece on plug.

5 Carry out additional alignment of workpiece as necessary.

Boring through holes

1 Engage boring head in machine spindle

2 Select boring bar
Select a bar long enough to cover depth of bore, and give approximately 1 in. (25 mm) clearance between boring head and top of workpiece when at full depth.

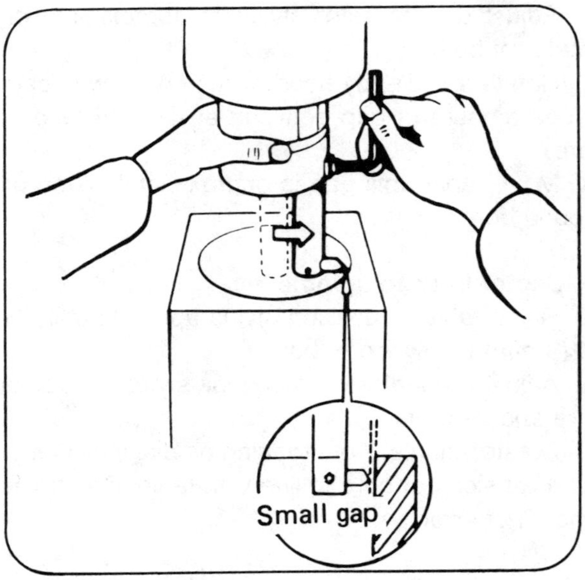

Note:
The tool hole may be positioned at 45° or 90° to the bar.

3 Mount boring bar
(a) Mount bar in boring head, and ensure that tool hole is directly in line, and centered to, feed slide index.
(b) Fully tighten retaining screw.

4 Engage tool in bar

Note:
Use the most rigid bar possible.

5 Set tool to hole diameter
(a) Position point of tool below top of bore.
(b) Move tool out until very small gap is between tool and bore.
(c) Clear tool over top of bore.

6 Take trial cut
(a) Select correct feed and speed.
(b) Set cut to leave approximately .010 in. (.25 mm) in bore and lock tool slide.

Note:
Twice the amount moved on some boring head indexes will be removed from the bore.
(c) Take cut approximately ⅛ in. (3 mm) down bore, using plenty of coolant.
(d) Clear tool from bore. Machine spindle remains centered to bore.
(e) Check bore diameter.

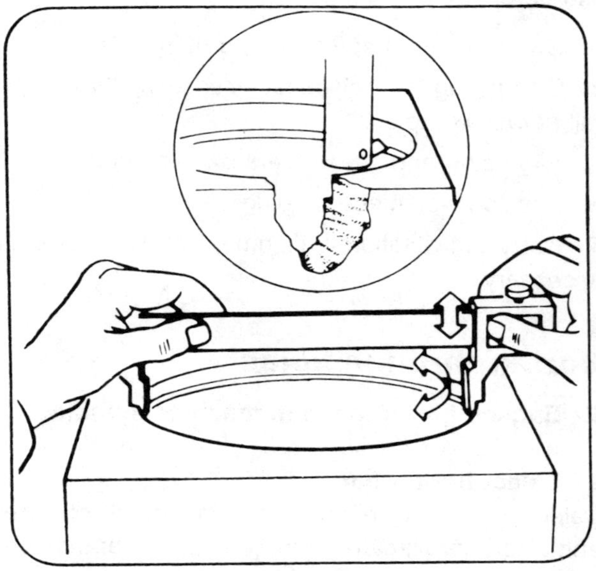

Note:
If it is necessary to move the workpiece, set table and saddle collars to zero for easy resetting, and **note direction of backlash**.
(f) Adjust cut to leave approximately .005 in. (.1 mm) for finishing cut, and lock tool slide.
(g) Adjust cut to leave approximately .004 to .005 in. (.1 mm) for finishing cut, and lock tool slide.

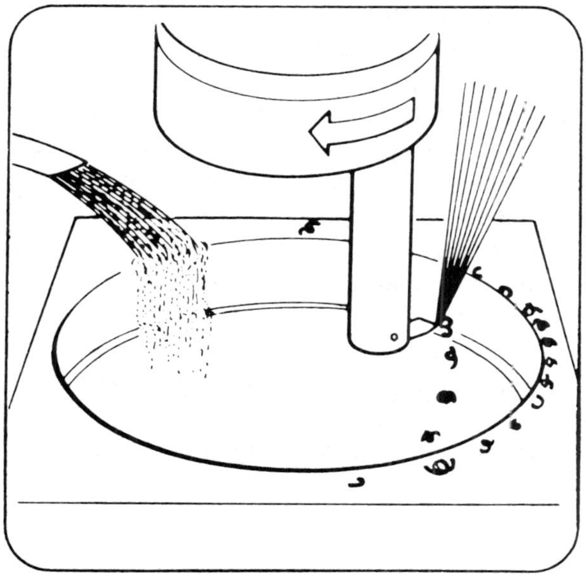

7 Take rough cut through

(a) Use plenty of coolant.

(b) Clean, bright chips should be removed. An even cutting sound denotes correct cutting action.

(c) Cutting sound ceases as cutter clears bottom of bore.

8 Check bore diameter and set finishing cut

(a) Clear tool from bore (spindle stopped).

(b) Check surface finish.

(c) Check bore diameter.

(d) Set finishing cut, allowing for deflection of tool during previous cut.

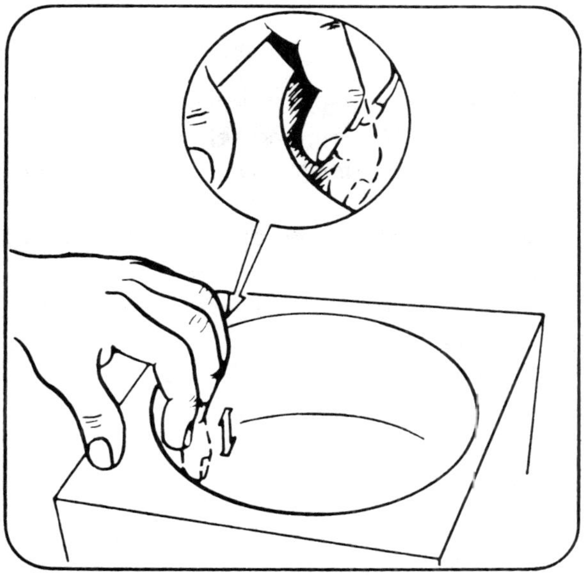

9 Take finishing cut

(a) Visually check amount of material being removed.

(b) Check bore diameter.

(c) Take additional cut(s) if required.

Note:

For bores longer than 3 in. (75 mm), a slower spindle speed tends to reduce cutter wear and chatter. The finishing cut taken through the bore twice tends to remove errors caused by tool deflection.

Boring holes to depth

A bore of a predetermined depth is first drilled out to leave, if possible, approximately 1⁄32 in. (.8 mm) for the boring operation. The depth may be squared out with a flat bottom drill.

1 Select boring bar

(a) Select bar long enough to cover full depth of bore.

(b) Ensure that tool hole is so positioned that cutting edge of tool will extend below end of bar.

Note:

A bar with a tool hole at 45° is normally used for blind bores. Alternatively a forged boring tool may be mounted directly into the boring head.

2 Select boring tool

(a) Select tool with cutting edge ground at appropriate angle, to form bottom face at 90° to sides of bore when mounted in bar.

(b) Ensure that bore is as rigid as possible.

3 Set tool to depth

(a) Move tool out, and position cutting edge over side of bore, projecting over face of workpiece.

(b) Raise table to bring cutting edge of tool within approximately 1/16 in. (1.5 mm) of face of workpiece.

(c) Insert feeler gage between tool edge and face of workpiece.

(d) Lower quill to lightly trap feeler.

(e) Zero collar (allowing for feeler thickness).

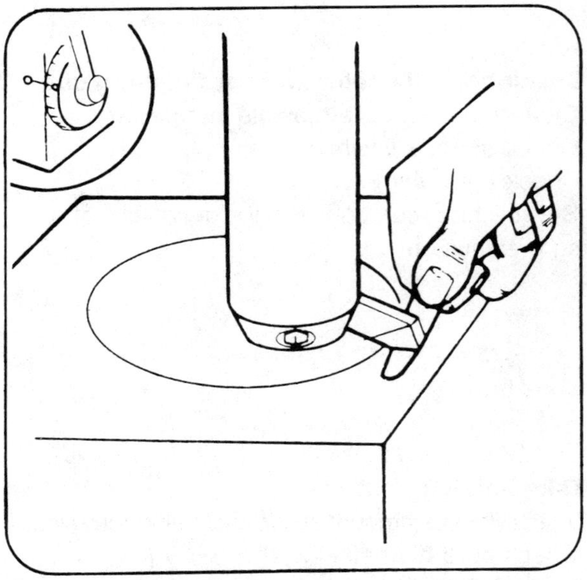

4 Set feed stop

(a) Move tool back to give small gap between tool and bore.

(b) Position quill at zero setting, at top of bore.

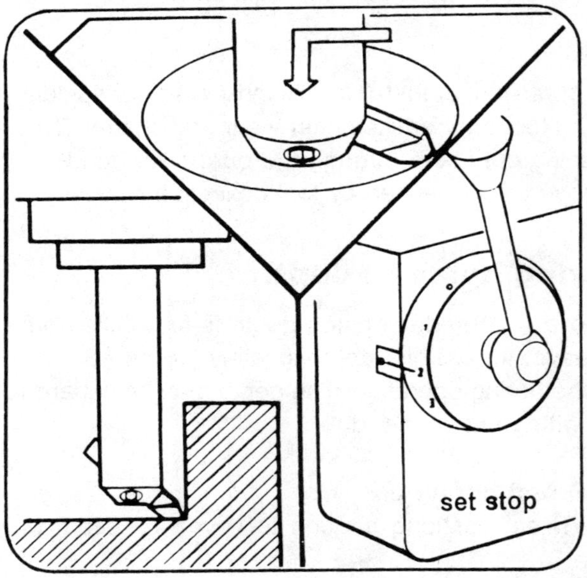

(c) Wind quill down required amount less approximately .010 in. (.25 mm), and set feed stop.

(d) Raise quill approximately .050 in. (1.25 mm) and start machine spindle.

(e) Engage quill feed lever, and allow quill to feed down to stop.

(f) Visually check that tool is giving good corner blend at bottom of bore.

(g) Adjust quill stop if necessary.

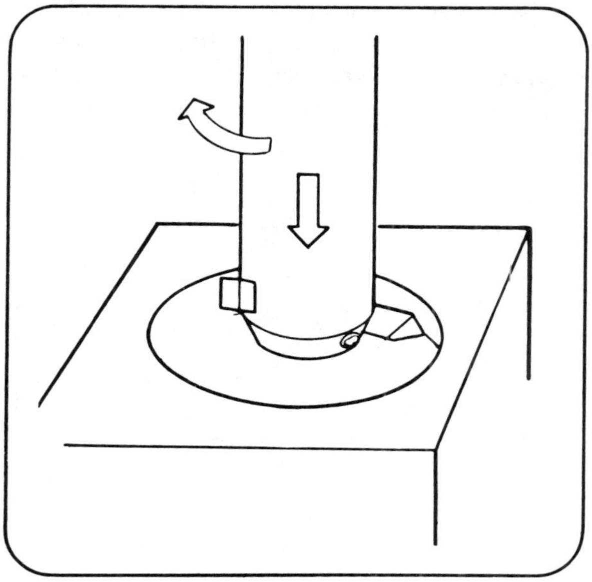

Note:
If the machine is not equipped with a feed stop, the quill should be fed down to within approximately .010 in. (.25 mm) of full depth. Knock out the feed lever at this point, and hand feed to full depth.
(h) Repeat steps 5, 6, 7, and 8, as for "Boring through holes."

Note:
If a boring head with a self-acting cross slide is available, the head has a built-in drive that is set in motion by the braking of a knurled ring. An adjustable stop is provided, whereby the outward travel of the cutter slide may be set to the finished bore size. The depth is first set to within approximately .003 in. (.07 mm) of finished size, and the bottom face of the bore is finish faced after the bore has been finished to size.

Checking distance between bores

1 Select plug gages
(a) Select plug gages approximately sized for bores to be checked.
(b) Clean and deburr bores, and clean plug gages.
(c) Insert gages in bores, ensuring good push fit.
(d) Allow plug to extend approximately ½ in. (12 mm) above face of workpiece.

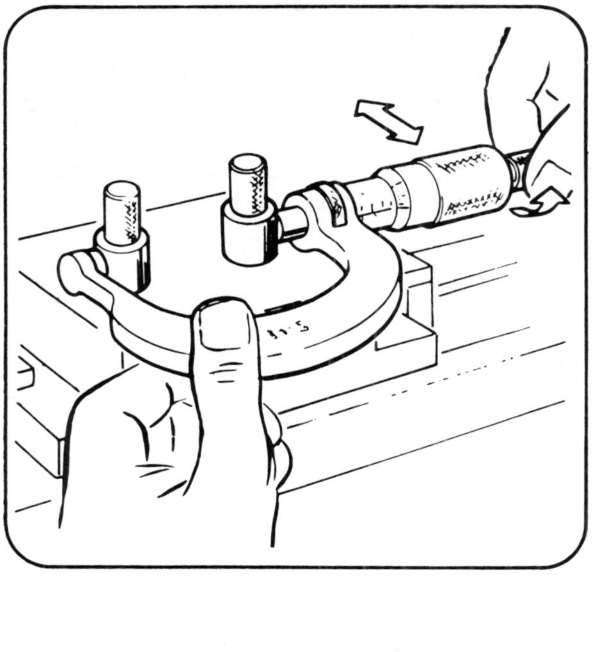

2 Measure distance between bore centers
(a) Measure distance over plug gages using suitable micrometer.
(b) Calculate required distance.

Example:

$$\text{distance between bore centers} = \text{micrometer reading} - \frac{\text{sum of diameters of plug gages}}{2}$$

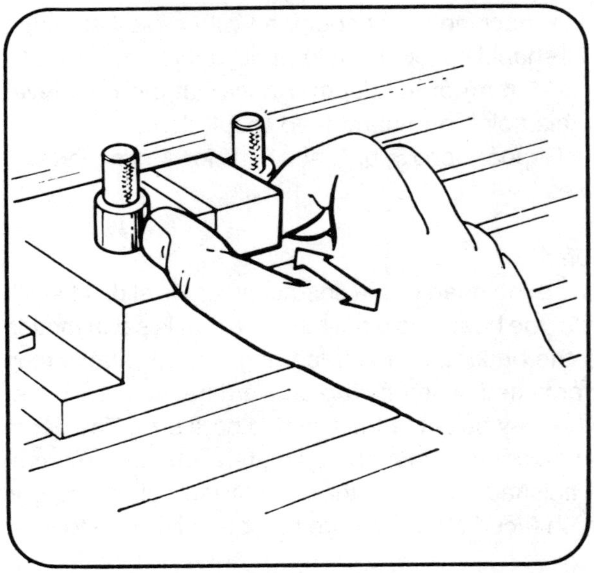

Note:
It may be more convenient, particularly on short bore centers, to use gage blocks and check between the plug gages. Alternatively, a caliper may be employed if the distance between the bore centers is great.

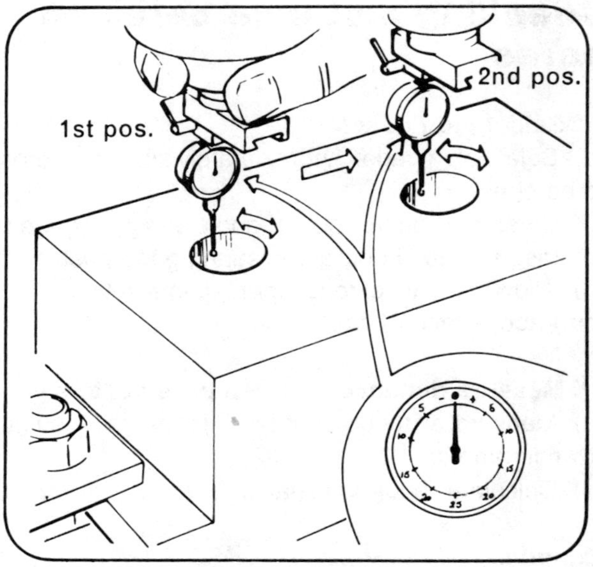

ALTERNATIVELY

1 Check first bore
(a) Mount dial indicator in machine spindle.
(b) Take machine out of gear.
(c) Engage dial indicator stylus in bore.
(d) Take reading around bore (first position).
(e) Clear stylus from bore.

Note:
The machine should be set in the correct centered position from the boring operation.

2 Set required distance
(a) Attach dial indicator to suitable point on stationary machine casting.
(b) Attach suitable stop to slide at front of table.
(c) Build up gage blocks to required distance between bores.
(d) Place gage blocks between stop and stylus of dial indicator.

(e) Tighten stop, and set dial indicator at zero.
(f) Mark direction of free movement of table (backlash).

3 Check distance between bores

(a) Remove gage blocks and move table to touch stop lightly on stylus of dial indicator.
(b) Adjust carefully to bring dial indicator to zero, ensuring that completed setting is in the direction marked for free movement (second position in illustration for step 1).
(c) Check around second bore with arbor-mounted dial indicator and note error reading.

Checking distance between bores and faces

1 Select plug gage

(a) Select plug gage appropriately sized for bore to be checked.
(b) Clean and deburr bore, and clean plug gage. Insert plug gage in bore, ensuring good push fit.
(c) Allow approximately ½ in. (12 mm) of plug to extend from front of bore.

2 Measure distance between bore and face

Using height gage measure:

(a) Distance from table to underside of plug (see diagram A).

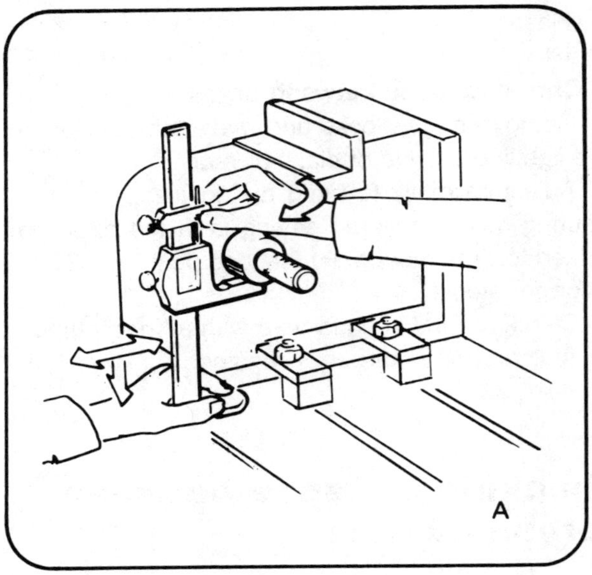

(b) Distance from table to face, using fingers (see diagram B).

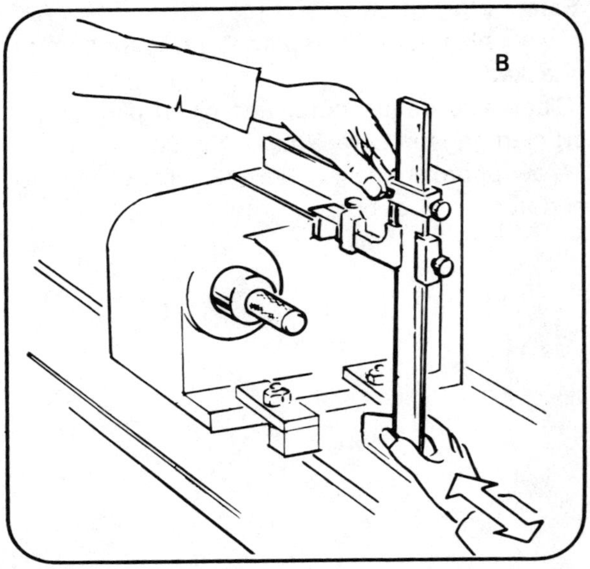

$$\text{distance between bore center and face} = \text{difference in gage readings} - \frac{\text{bore diameter}}{2}$$

Other boring bar operations

Counterboring

A single-point tool is used for counterbores of large diameter, when close tolerances are required, or when the counterbore is so positioned that it is more convenient to machine by this method. The through hole is finish bored, and the application is then identical to that of boring holes to a predetermined depth. It should be noted that the centered position of the machine spindle is maintained from the first operation.

Recessing

Recessing can be carried out only if the boring head is equipped with a built-in drive. The bore is first finished to size. The appropriate tool is then mounted in the bar and set to the required position in the bore. The tool is ground to the required width, and the cross drive of the head is used to feed the tool to the required depth.

1 Select boring tool

Select suitable tool, of appropriate width for recess.

2 Set tool to position of recess

(a) Move tool out, and position cutting edge over side of bore, projecting over top face of workpiece.
(b) Raise table to bring edge of tool within $1/16$ in. (1.5 mm) of workpiece.
(c) Insert feeler gage between tool edge and top of workpiece.
(d) Lower quill to lightly trap feeler, and zero collar (allowing for feeler thickness).
(e) Move tool back to clear workpiece.

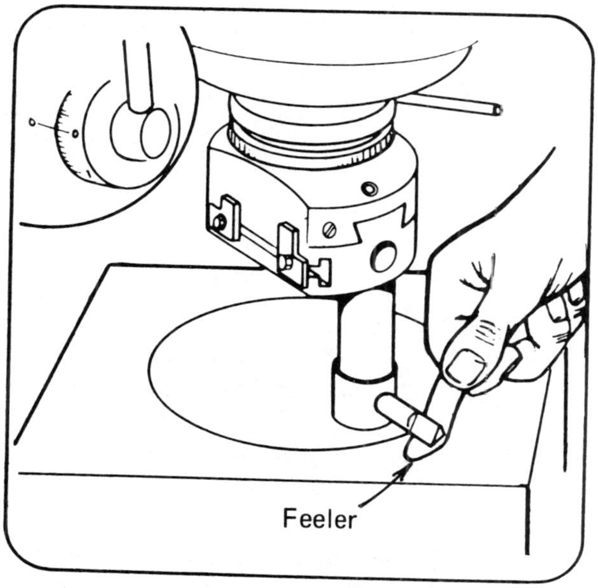

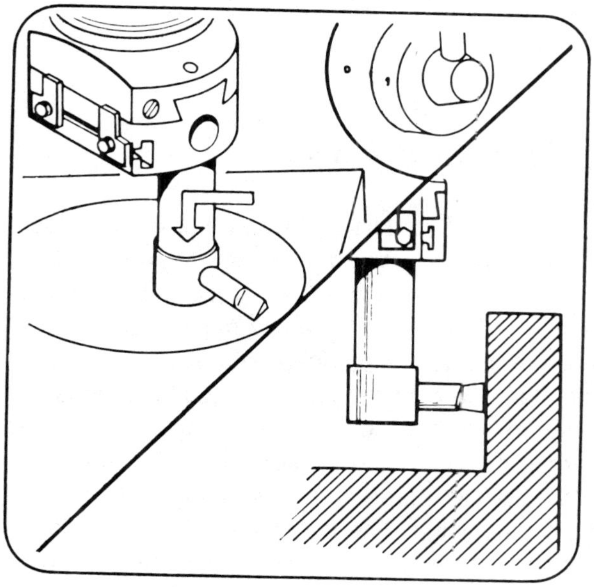

3 Set tool to cutting position

Drop quill, using collar, to required position. Reset collar at zero and lock quill.

Note:
distance to required position =

$$\frac{\text{width of tool}}{2} + \begin{array}{l}\text{distance to center of recess}\\ \text{from top of bore}\end{array}$$

4 Cut recess

(a) Take machine out of gear.
(b) Rotate graduated dial on head until very small gap is obtained between bore and tool.

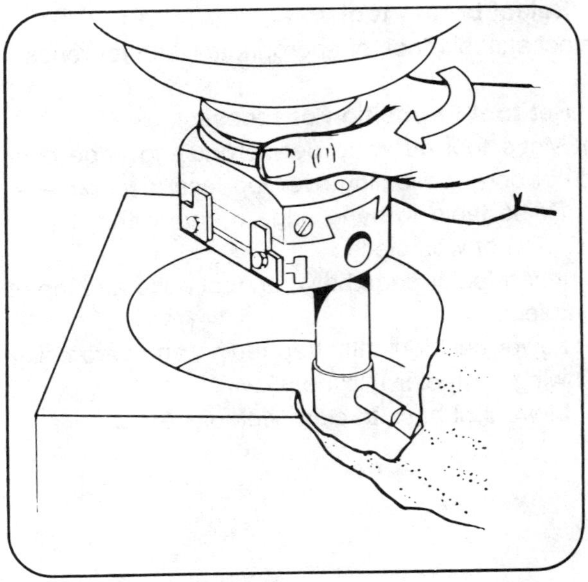

(c) Continue to rotate dial very gently, turning head by hand, at intervals, until face of tool just scrapes bore.

(d) Engage gear and start machine.

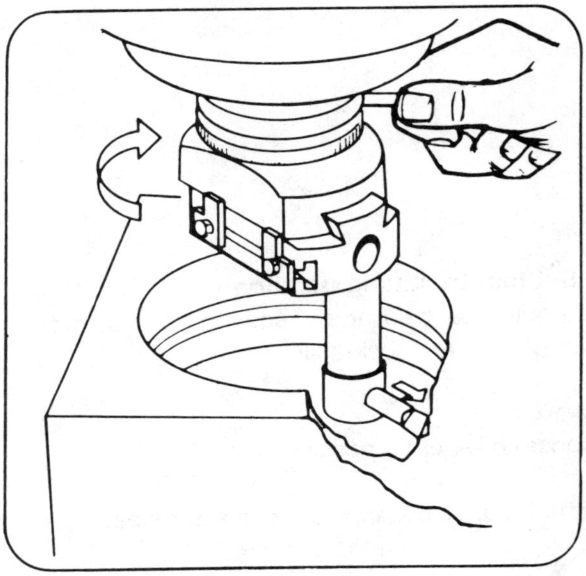

(e) Advance tool for depth of cut required.

(f) Stop machine and withdraw tool from recess and hole.

(g) Check diameter and position of recess.

Note:

On some boring heads twice the amount moved on the graduated ring will be removed from the diameter of the recess.

Milling castings

Castings are of various shapes and sizes. It is customary, before beginning any machining operation, to lay out castings to check for sufficiency of stock. The drawing must be carefully studied and a sequence of operations planned before any machining takes place, to ensure that as many faces, bosses, and so on as possible may be machined while the casting is set in one particular position. Careful note should also be made of any restricted areas that require machining. The casting should be so positioned that these areas may be machined without hitting other parts of the casting or workholding.

In some instances, for ease of workholding or setting, it is necessary to machine faces or bosses that need not be machined according to the drawing. A minimum of material is removed to give a flat face.

The largest face or base of the casting is normally used as a reference face and is machined first. If the casting has flanges or sides around the base, these are machined while the casting is set up in the position for machining the base. Drilled and reamed holes or bores are also machined directly after machining faces and so on. The clamped position of the casting is maintained and the holes positioned and bored through the machined faces before proceeding to the next setting or moving of the casting.

In the event of large blow holes, or a number of small blow holes, appearing on faces or in bores of the casting, after cleaning up or on reaching finished size, a supervisor should be consulted before further operations are carried out.

Procedure for castings

1 Mount workpiece

(a) Select suitable workholding method to machine base of workpiece.

(b) Measure workpiece to ensure that there is sufficient machining allowance.

(c) Position workpiece so that rough base or marked reference line is parallel or horizontal to machine table. Use shims or jacks if necessary.

(d) Check with surface gage from table or vertical column of machine to base, or layout if applicable.

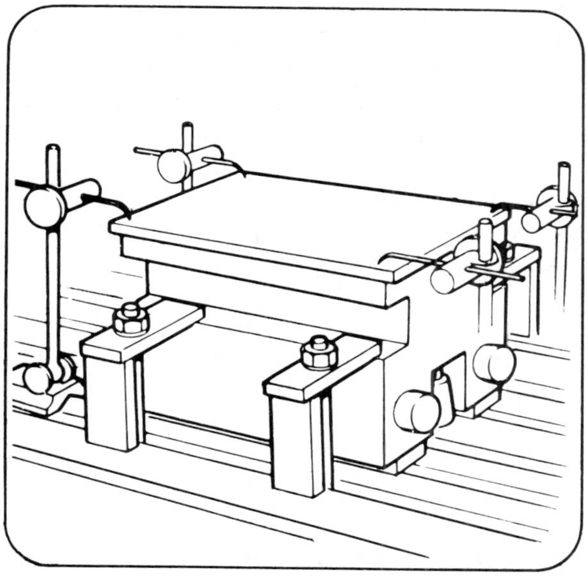

2 Clamp workpiece

(a) Select suitable clamps and shims.

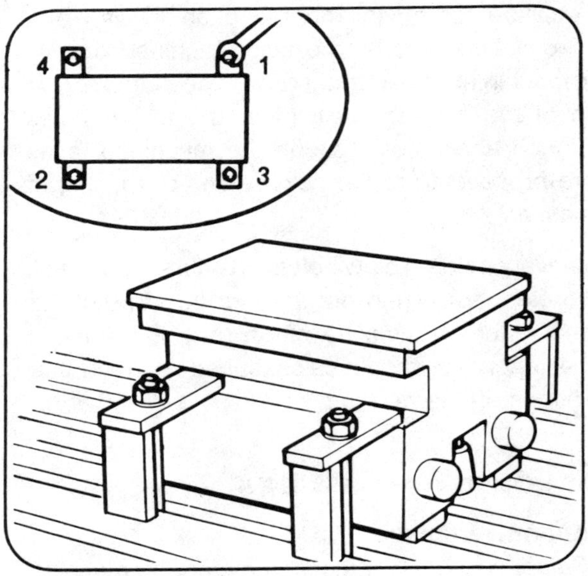

(b) Clamp workpiece, ensuring that whole face may be machined without hitting clamps or workholding, and that clamps are positioned over a solid member, such as jacks or shims. Clamps should never be placed over frail or thin walls of a casting, even when supported by jacks. A dial indicator may be used to detect workpiece distortion.

(c) Apply even, firm pressure to clamp bolts, ensuring that no distortion occurs.

(d) Recheck with surface gage to ensure that workpiece has not moved.

3 Select cutter

Select suitable face mill, having larger diameter than face to be machined.

4 Machine reference face

(a) Position table to cover face of workpiece with cutter.

(b) Raise table and take sufficient depth of cut to get below scale on workpiece.

(c) Take additional roughing cuts as necessary, leaving small amount for final finishing cut.

(d) Raise table required amount to split marked line, if machining to layouts.

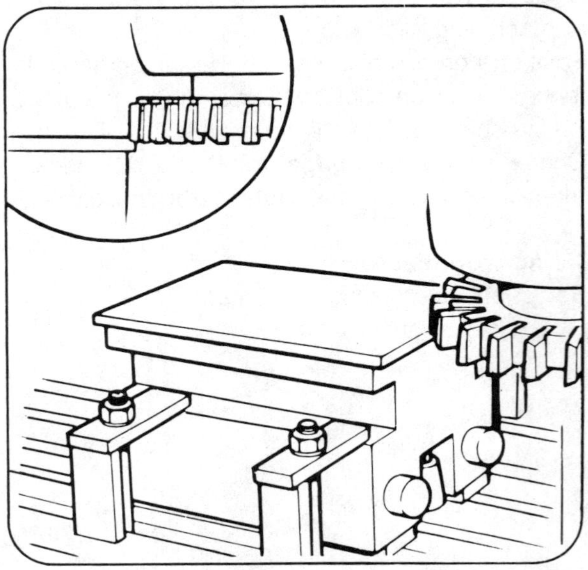

5 Mill sides of base

(a) Remove face mill and mount suitable end mill in machine spindle.

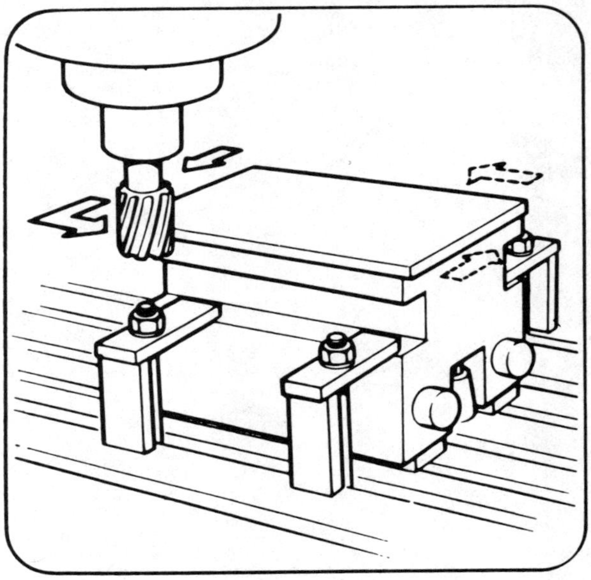

(b) Mill around the four sides or flanges of base while workpiece is set in position used for milling base.

Note:
This operation is applicable only if the workpiece has flanges or sides around the base.

6 Mount workpiece on base

(a) Clean machine table and machined base of workpiece.

(b) Clamp workpiece on table, aligning one side of base to keys mounted in table slot at rear of machine.

(c) Position workpiece to enable as many faces, bosses, and so on as possible to be machined without moving workpiece.

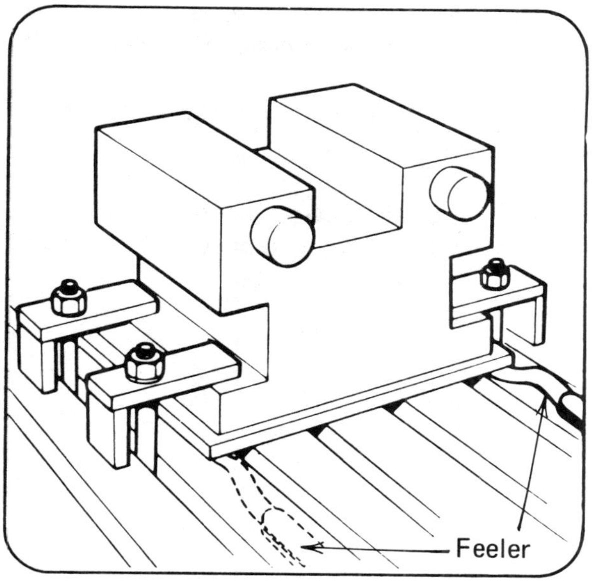

(d) Check around base with feeler gage to ensure that it is seating correctly on machine table.

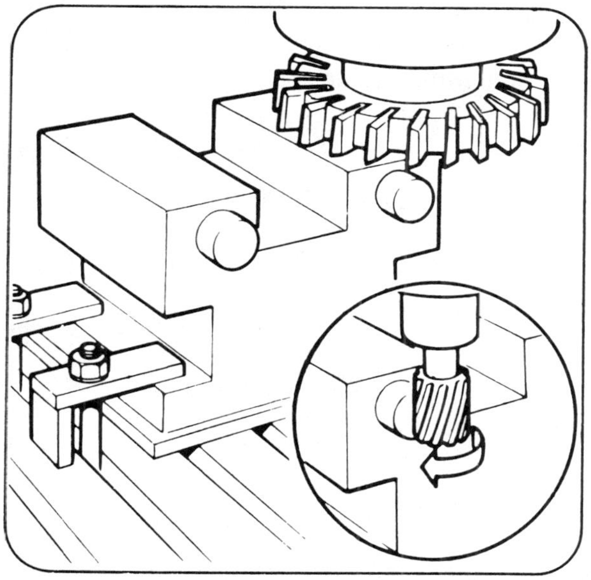

7 Machine top faces, bosses, etc.
(a) Mount suitable cutter in machine spindle, and machine as many faces and bosses as possible. Care should be exercised when machining any restricted areas.
(b) Bore any holes required while workpiece is set in this position.

Note:
It may not be possible to machine *all* faces, bosses, and so on with the workpiece clamped to the machine table. In such cases, the workpiece is mounted on an angle plate or similar workholding method, clamping the base or reference face to the plate.

Workpieces may be machined on horizontal machines in the same manner, the base being mounted at the rear of machine table, and a face mill mounted in machine spindle. Subsequent faces, bosses, and so on are machined with a side and face cutter or slab mill.

Facing to length: Vertical mill

1 Select cutter
Select an end mill long enough to cover face of workpiece.

2 Mount workpiece
(a) Select suitable pair of parallels, and mount workpiece with top face approximately 1/16 in. (1 5 mm) above vise jaws.

(b) Ensure that workpiece extends far enough from end of vise to remove material required without hitting vise jaws.
(c) Mount so that direction of cut pressure is toward fixed vise jaw.

3 Position cutter to workpiece

(a) Raise table to bring end of cutter below bottom face of workpiece.

(b) Adjust table to touch cutter lightly to workpiece, (cutter revolving).
(c) Set table collar at zero, and move cross slide to clear cutter from workpiece.

4 Face one end

(a) Check length of workpiece with ruler, and note amount of material to be removed to bring to finished size.
(b) Move longitudinal slide required amount to remove approximately half the excess material. Lock slide in position for cut.
(c) Power feed cross slide, using plenty of coolant.
(d) Check face with square.

5 Finish to length

(a) Remove workpiece from vise, deburr, and clean vise jaws and parallels.
(b) Turn workpiece and mount as before, ensuring that cutter will not hit vise jaws at finished length.
(c) Touch cutter to workpiece, and remove material to bring to finished length plus approximately .020 in. (.5 mm).

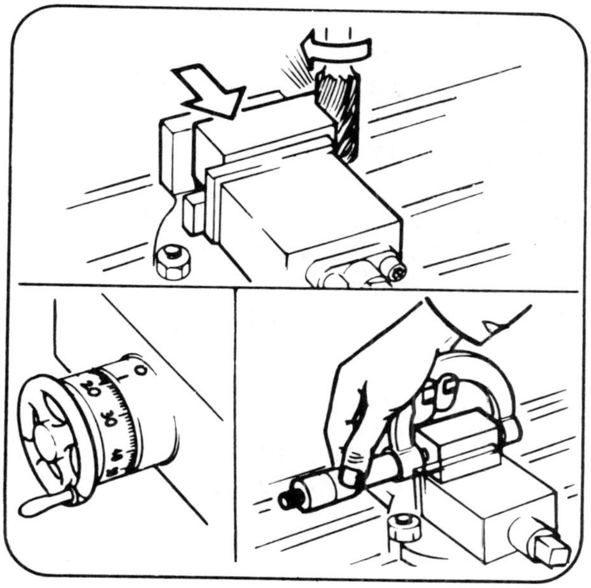

(c) Check with micrometer, and remove material to bring to finished length.
(e) Recheck before removing workpiece from vise.

Machining with workpiece vertical in a vise

Short workpieces may be mounted vertically in the vise and set with a square for the first face. Only workpieces that will not extend more than approximately 2 in. (50 mm) above the vise jaws, before being faced to length, should be machined in this way.

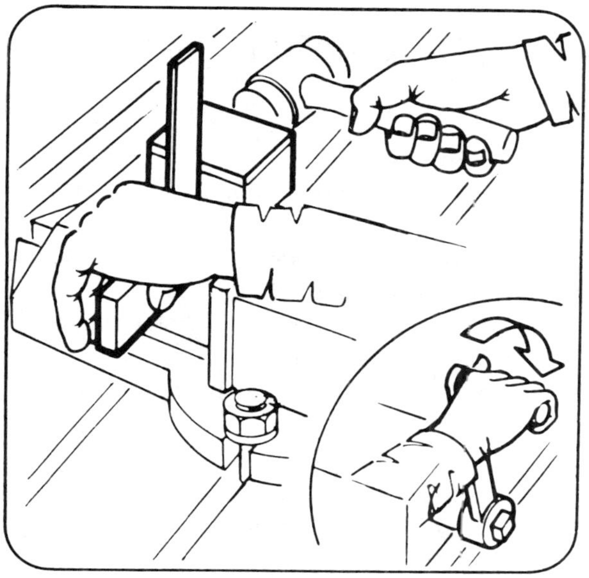

1 Mount workpiece
(a) Visually align workpiece in vertical plane and hand tighten vise.

(b) Place a square on bottom face of vise, blade along vertical face of workpiece.
(c) Tap workpiece gently with soft hammer, to align workpiece to square and fully tighten vise.

2 Select cutter
Select a shell end mill, or face mill, of sufficient diameter to cover face of workpiece.

3 Position cutter to workpiece
(a) Move table, to bring cutter over workpiece.
(b) Raise table, cutter revolving, until cutter lightly touches workpiece.
(c) Raise table minimum amount to clean up face of workpiece, and take cut.

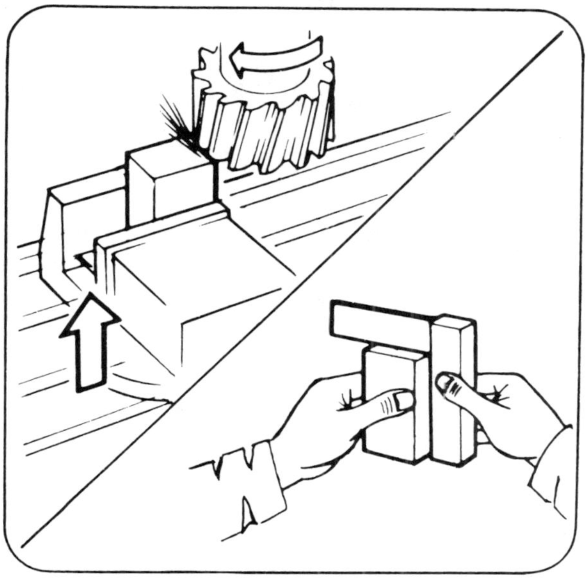

(d) Check workpiece with square.

4 Finish to length
(a) Remove workpiece from vise, deburr, and clean vise jaws and parallels.

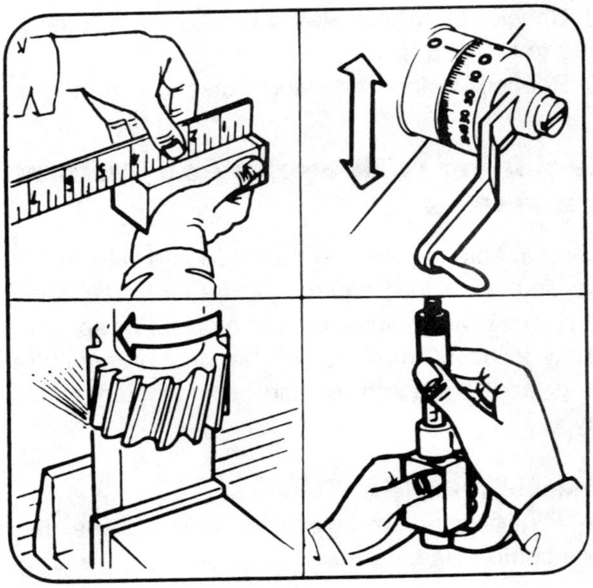

(b) Check workpiece with ruler and note amount of material to be removed.
(c) Turn workpiece and mount in vise; seat on machined end.
(d) Touch cutter lightly to workpiece, and set vertical collar at zero.
(e) Remove material to leave approximately .030 in. (.8 mm) on finished size, using plenty of coolant.
(f) Remove workpiece and check with micrometer.
(g) Remount and remove material to bring to finished size.

Note:
On very short workpieces, parallels may be required to bring the face of the workpiece above the vise jaws.

Facing to length: Horizontal mill

1 Mount workpiece

(a) Select a suitable pair of parallels on which to mount workpiece.
(b) Ensure that end of workpiece extends far enough from vise to remove material required.
(c) Position workpiece so that direction of cut pressure is toward fixed vise jaw.
(d) Tighten vise, and tap workpiece down until parallels are trapped.

2 Select cutter

Select side and face cutter having diameter large enough to cut full face of workpiece without arbor collars hitting top of workpiece.

3 Position cutter to workpiece

(a) Bring face of cutter below bottom edge of workpiece and close to it.
(b) Start machine, lightly touch cutter to workpiece, and zero collar.
(c) Move table, to clear cutter from workpiece.

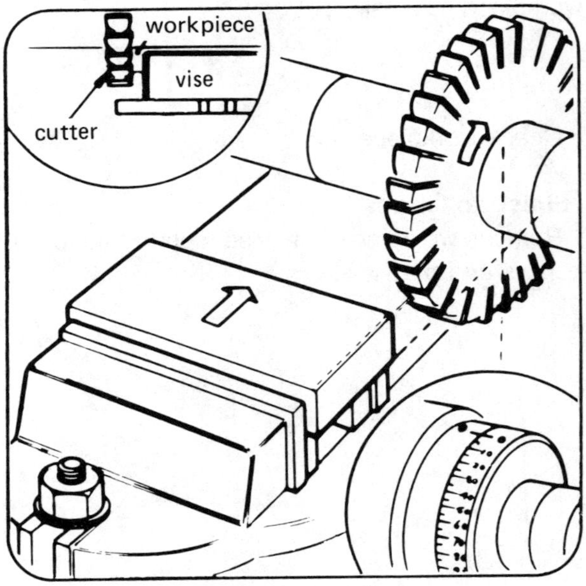

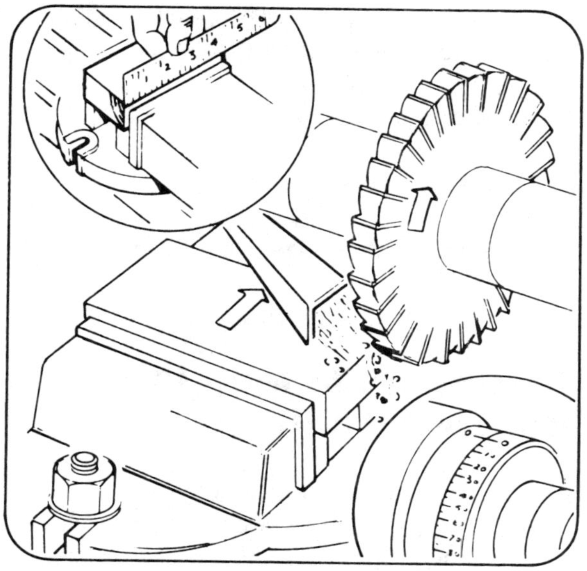

4 Face one end of workpiece

(a) Check length of workpiece with ruler. Note amount of material to be removed.
(b) Move cross slide required distance to remove most of the required amount, and lock slide.
(c) Power feed table, using plenty of coolant when cutting.
(d) Check face with square, and visually check surface finish with cutter stopped and clear of workpiece.

5 Face to finished size

(a) Check with micrometer, and calculate finishing cut required.
(b) Move saddle required distance and take finishing cut, using plenty of coolant.
(c) Check finished size and deburr.

Note:
1 For workpieces with excessive amounts of material on the length, two or more cuts may have to be taken to bring to finished size and to obtain a good surface finish.
2 If both ends require machining, repeat the technique described above for the opposite end.

Facing to length with face mill

1 Select cutter

Select face mill of slightly larger diameter than face of workpiece.

2 Mount face mill on machine spindle

3 Mount workpiece

4 Position cutter to workpiece

(a) Visually set center of cutter approximately 1/8 in. (3 mm) below top surface of workpiece.
(b) Bring face of cutter close to workpiece.

(c) Start machine, ensuring that cut is toward fixed vise jaws and that cutter will cut down. Lightly touch cutter to workpiece and zero collar.

5 Face one end

6 Deburr workpiece

7 Face to finished size

Cutting off to length: Horizontal mill

Bar or plate may be cut to length with a slitting saw mounted on the machine arbor. The workpiece may be held in the vise or clamped to the machine table. Leave a minimum amount of material on the workpiece for removal in subsequent operations.

Workpiece held in vise

1 Position cutter to workpiece

(a) Select suitable pair of parallels on which to mount workpiece.
(b) Position bottom of cutter just below bottom face of workpiece.
(c) Bring cutter close to workpiece, ensuring that direction of cut pressure is toward fixed vise jaw.
(d) Position saw close to end of vise jaw. Lock slide.

2 Set for length

(a) Adjust workpiece as shown in diagram to cut off required length.

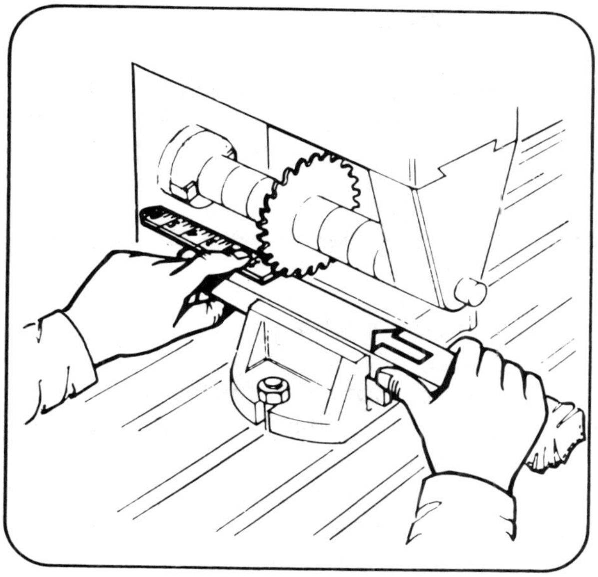

(b) Ensure that end of workpiece will not hit driving dogs on machine spindle.
(c) Tighten vise, and tap workpiece with soft hammer to seat tight on parallels.

Note:
Where several pieces need to be cut, stops can be used on the machine to avoid repetitive measurements.

Workpiece clamped to table

1 Position workpiece and set for length

(a) Clamp suitable parallel square across table. Position workpiece as shown in diagram and clamp to table.

(b) Set length by method described for workpiece held in vise, ensuring that diameter of saw is sufficient for arbor collars to clear top of bolts and clamps as indicated in diagram.
(c) Position saw in a table slot, to enable workpiece to be sawn through without hitting face of table.
(d) Ensure that direction of cut pressure is toward parallel.

2 Cut off to length

(a) Start machine and hand feed table gently, to allow slitting saw to feed into workpiece for approximately 1/8 in. (3 mm).
(b) Engage power feed, and allow cutter to feed through workpiece using plenty of coolant.
(c) Visually check chips to ensure that correct feed has been selected.
(d) Disengage feed when cutter is within approximately 1/8 in. (3 mm) of cutting through bar. (Piece being cut off will vibrate.)
(e) Hand feed remaining distance gently to allow piece to drop from end of bar.
(f) Deburr end of bar before resetting for next length.

Safety

Attach a piece of rag or paper to the end of the bar extending from the vise. Always wear safety glasses when using a saw.

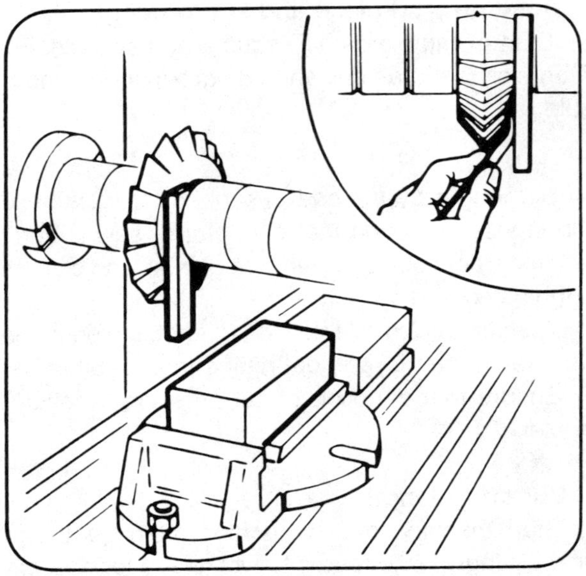

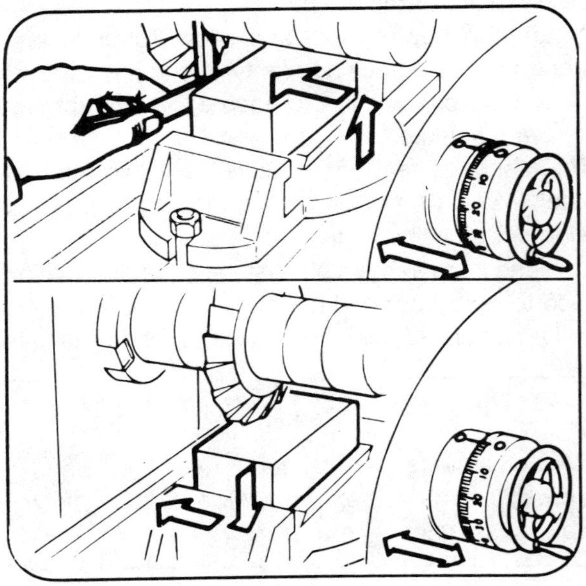

Milling vees

Vees are milled with an arbor-mounted angle cutter. Alternatively, the workpiece may be laid out, and set at the required angle, in the vise.

Milling vees with an equal-angle cutter

1 Mount cutter on arbor

2 Prepare arbor for cutter positioning

(a) Loosen arbor locking nut, and insert small parallel bar between cutter and collar.
(b) Ensure that parallel extends approximately 1/4 in. (6 mm) below diameter of cutter.
(c) Tighten arbor locking nut.
(d) Check between cutter boss and parallel with feeler gage, to ensure good seat.

Note:
The cutter should be checked to ensure that the angles are ground equally about the center.

3 Position cutter to workpiece

(a) Bring parallel close to end of workpiece.

(b) Insert feeler gage between parallel and workpiece, and gently move across slide to lightly trap it.
(c) Zero collar. Drop table to clear parallel from workpiece.
(d) Move cross slide required distance to position center of cutter. Reset collar on zero.

$$\text{distance of center of vee from end of workpiece} + \frac{\text{width of cutter}}{2} + \text{thickness of parallel} + \text{thickness of feeler gage} = \text{required distance}$$

(e) Loosen arbor locking nut, remove parallel, and retighten.

4 Set cutter for depth

(a) Bring cutter over face of workpiece.

(b) Raise table (cutter revolving) to bring cutter so that points of cutter lightly mark workpiece.
(c) Zero collar.

5 Cut vee

(a) Remove material in even cuts, until about ¼ in. (6 mm) depth has been reached.
(b) Use plenty of coolant, remove chips with brush (cutter stopped).
(c) Reduce depth of cut after ¼ in. (6 mm) depth has been reached.

Milling vees with work at the required angle

Vees of 90° or more inclusive angle may be cut using the following method. If the inclusive angle required is less than 90° an appropriate angled cutter must be used.

1 Lay out workpiece

Apply layout dye to workpiece. Mark centerline and width of vee on top face, and depth of vee and correct angle on each end.

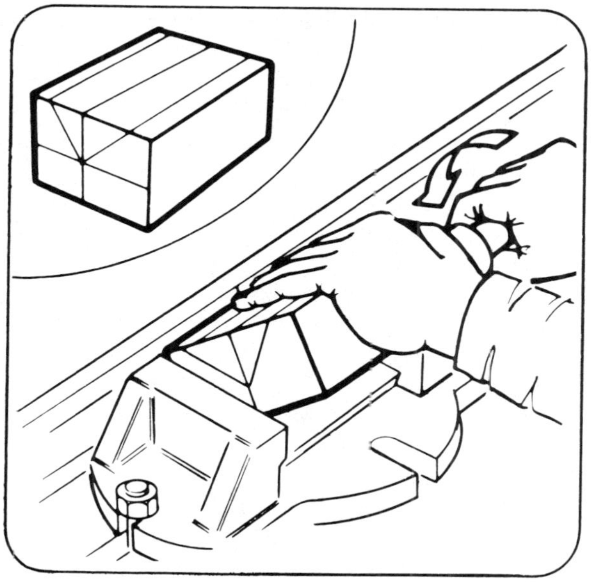

2 Mount workpiece

(a) Set workpiece in vise. Visually align one marked line of vee in horizontal plane, using top of vise as guide.
(b) Ensure that workpiece is as low as possible in vise, with marked horizontal line slightly above vise jaw.
(c) Hand tighten vise to hold workpiece in position.

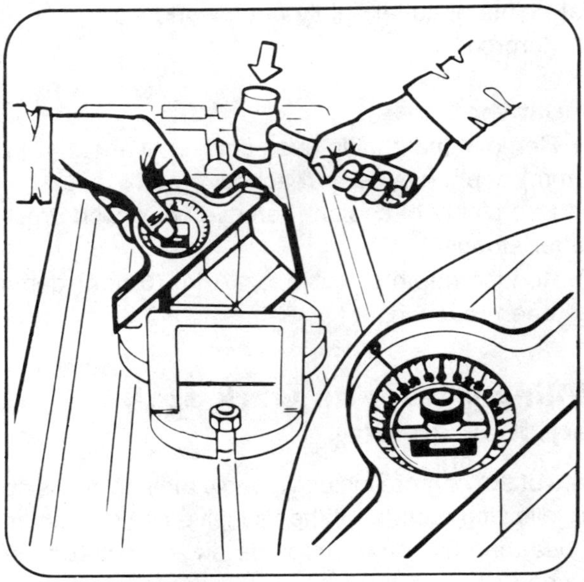

3 Set workpiece at required angle

Set as shown in diagram, using protractor or clinometer.

Note:
When using bubble instruments, first check the table level and, where appropriate, take observed errors into account when setting the workpiece.

4 Select cutter

Select a side and face cutter having diameter large enough to cut full depth of vee, without arbor collars hitting workpiece.

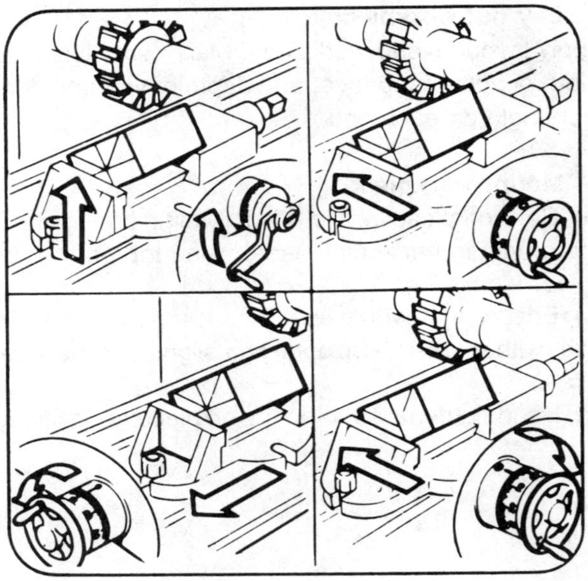

5 Cut vee

(a) Bring face of cutter above horizontal line of workpiece.
(b) Let cutter just touch workpiece. Zero collar.
(c) Move table to clear cutter from workpiece, and move to cut with cross slide.
(d) Remove material in even cuts, until side of cutter is close to vertical marked line.

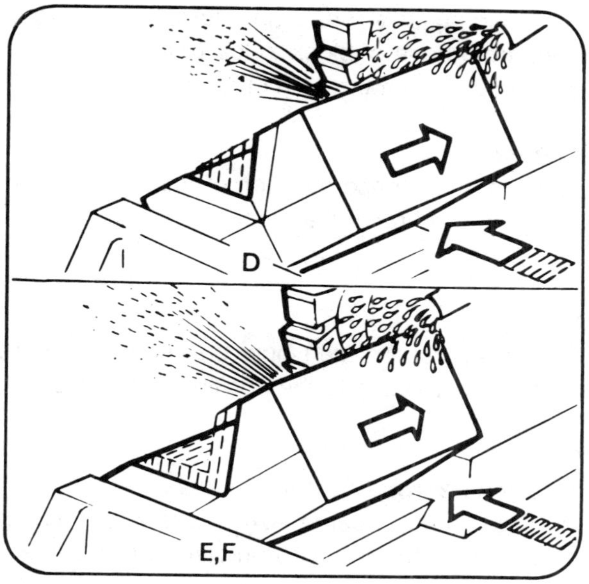

(e) Return table to starting point. Gently raise table so that face of cutter just splits horizontal line on workpiece.

(f) Remove material in even cuts as before, until side of cutter splits vertical line on workpiece.

Note:

1 The finishing cut will aid the surface finish obtained. If the material is removed in one cut, the pressure will tend to force the face of the cutter down, and it will cut below the horizontal line of the vee.

2 When the included angle is greater than 90°, the angles of the vee should be set and machined individually. The position of the vee must first be laid out on the workpiece.

Checking vees

1 Check symmetry of vee with centerline

(a) Select suitable roller, having diameter such that roller makes contact about midway down face of vee.

(b) Place roller lengthways in vee, and slightly extending over end of vee.

(c) Clamp roller in vee.

(d) Turn workpiece on its side and set dial indicator at zero on top of roller.

(e) Turn workpiece 180° and check for zero on top of roller.

2 Check depth of vee

(a) Select suitable diameter roller and mount in vee block as shown.

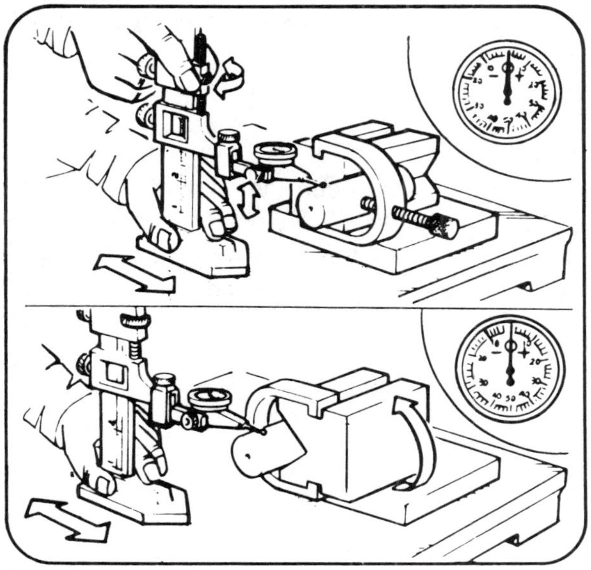

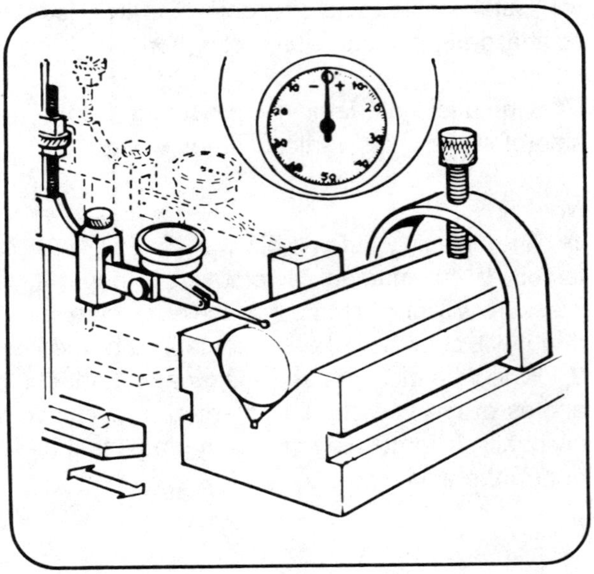

(b) Using suitable measuring device, measure distance from base of block to top of roller.
(c) Calculate distance to center of roller.

Example:
distance = distance over top of roller $-\dfrac{\text{roller diameter}}{2}$

Note:
This measurement can now be used to calculate dimensions relating to the position of the vee in the block.

3 Check across center of vees
(a) Select pair of same size rollers whose centers will be below top face of workpiece when seated in vees.
(b) Seat rollers in vees, ends slightly extending from end of workpiece.
(c) Note distance between centers of vees from drawings, and add diameter of one roller.
(d) Select caliper, and set to dimension required.
(e) Place hand across top of rollers and hold in position with downward pressure.
(f) Place caliper on outside of rollers, applying up-and-down, very slightly rotary movement to obtain a good feel.
(g) Adjust if necessary.

Note:
If the distance between the centers of the vees is small, gage blocks may be used between the rollers. Subtract the diameter of one roller from the vee centers when applying this method.

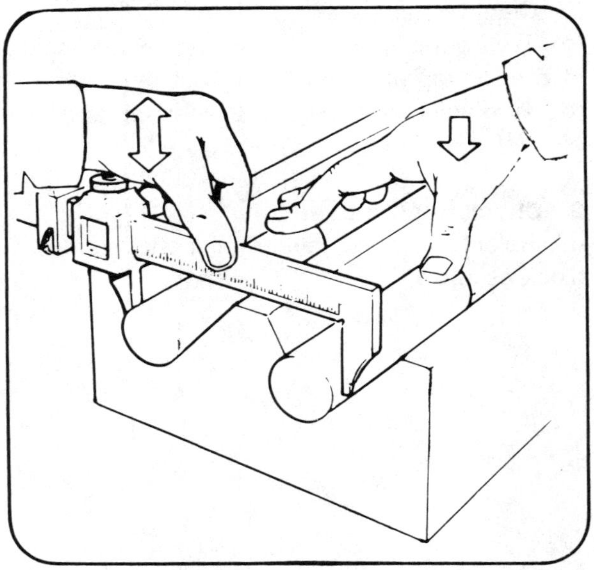

Review questions

1 Describe and demonstrate climb and conventional milling.

2 What are the dangers in climb milling on a conventional milling machine?

3 What are the advantages of climb milling?

4 Describe and demonstrate the correct procedure for milling a block square.

5 Describe and demonstrate procedures for angular milling.

6 Describe and demonstrate procedures for keyway milling and measurement.

7 Describe and demonstrate procedures for T-slot milling.

8 How is RPM for drilling on the mill calculated?

9 What is the relation of speed to feed in reaming?

10 Describe and demonstrate procedures for tapping in the mill.

11 What precautions should be observed while tapping a blind hole?

12 What is spotfacing?

13 Describe and demonstrate boring in the mill.

14 Describe and demonstrate bore diameter measurement and measurement of distance between bores.

15 Describe general considerations for milling of castings.

16 Describe and demonstrate facing to length in the vertical and horizontal mill.

17 Describe milling of vees.

18 How are vees measured?

19 How is internal recessing or grooving done on the mill?

20 How may surface finish be checked?

UNIT TEN

Rotary Tables, Indexing/Dividing Heads, and Related Milling Operations

Using the rotary table

On all operations of this nature, the machine spindle and the rotary table axis must be aligned and the attachment set relative to the table movement. Centers of radii should be marked on the workpiece, or, if the centers cannot be marked, the path of the keyway or slot should be laid out before mounting on the rotary table.

Slots and keyways

1 Position workpiece

(a) Mount workpiece on parallels on rotary table and set center point of radius central to machine spindle.

(b) Visually align reference face square or parallel to vertical column of machine.

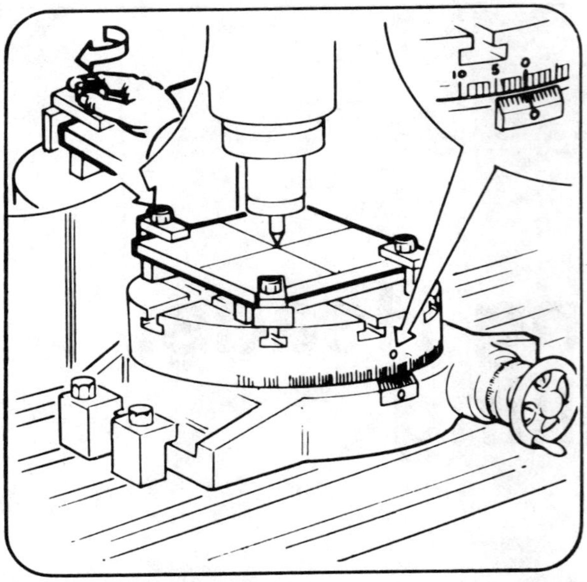

(c) Set scale on circumference of table, and vernier index dial on worm, at zero. Lock table slide.

(d) Assemble and attach bolts and clamps, and finger tighten. Position clamps so that cutter will not hit them when machining keyway or slot.

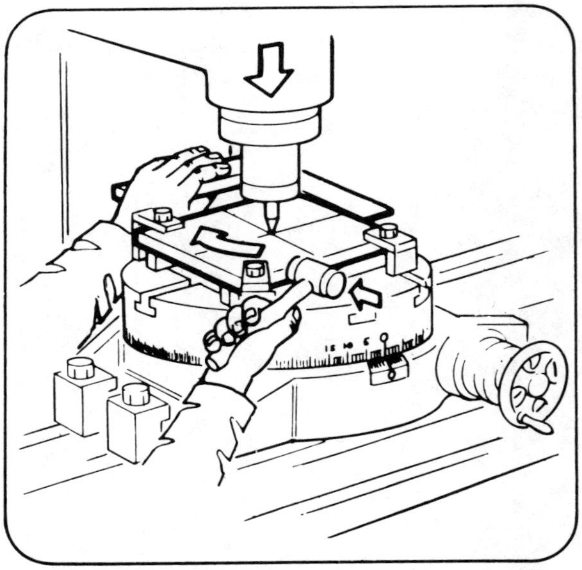

2 Align reference face

(a) Use square to align reference face (if applicable) square to vertical column of machine.

(b) Pressure on machine center will keep workpiece centered when setting is square.

(c) Fully tighten clamps, ensuring good seat on parallels.

Note:
If the reference face is parallel to the vertical column when the workpiece is in position for machining the keyway, index the rotary table around 90°, bringing the reference face square. Return to the original position after setting.

3 Select cutter

Select suitable end mill, having diameter correct for width of slot or keyway.

Note:
On wide slots, or keyways of close tolerances, the slot of keyway may have to be roughed out, leaving approximately .030 in. (.8 mm) on the width for finishing.

4 Position cutter to workpiece

(a) Move table slide required amount from center point for correct radius, and lock slide. Calculate cutter radius compensation if necessary.

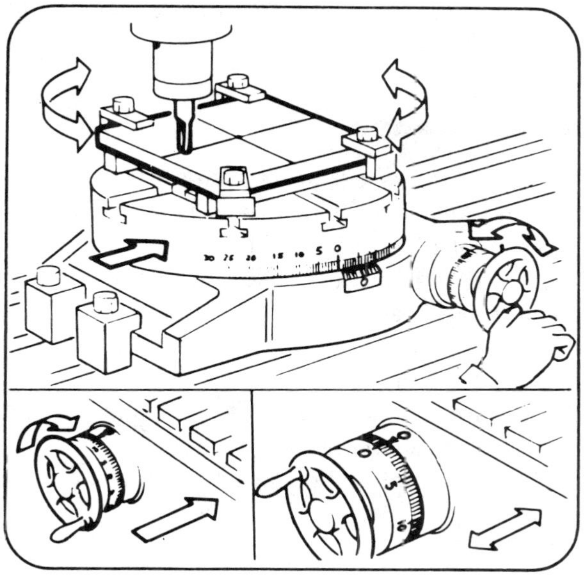

(b) Hand feed rotary table to starting point of keyway or slot, using the scale. Final setting is done by handwheel vernier collar.

(c) Reset handwheel collar at zero, and mark scale on circumference.

(d) Hand feed around to opposite end of keyway and mark scale on circumference and handwheel collar.

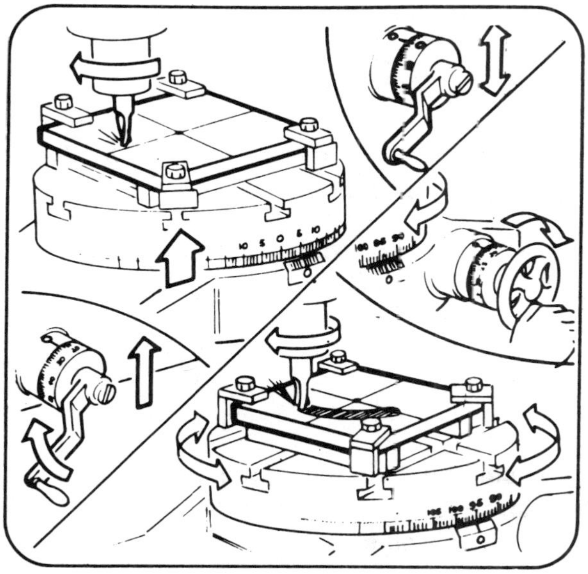

5 Set cutter for depth

(a) Start machine, and raise table gently until cutter lightly marks workpiece.

(b) Set vertical collar at zero.

(c) Index table around whole path of slot or keyway to ensure that position is correct.

6 Cut slot or keyway

(a) Raise table to remove material in even cuts, from each end of radius alternately, using plenty of coolant.

(b) Stop machine, and remove excess chips with brush if buildup becomes great.

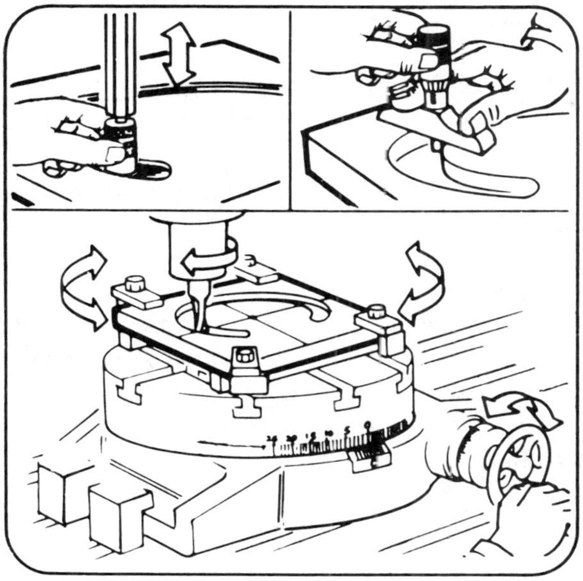

(c) Check width of keyway after depth of approximately 1/16 in. (1.5 mm) has been reached.

(d) Check depth of keyway with depth micrometer or caliper when within .020 in. (.5 mm) of full depth.

Safety

Be careful of your hands on the cutter when checking the keyway and removing chips with a brush.

Note:

For keyways and slots that have a long travel, the rotary table power feed may be used. Feed trip stops should be set approximately 3° short of the finishing point at each end. The path of the keyway is then completed by hand feed. Slots that pass through the workpiece are drilled or reamed out at each end of the radius before the path of the slot is cut.

Circular faces

Circular faces are set in the same manner as for slots and keyways. If a hole is positioned at the center point of the radius of the face, this is first drilled and reamed, and used as a reference for setting the radius.

1 Select cutter

(a) Select end mill, as short as possible, but long enough to cover whole face of workpiece.

(b) Check diameter of cutter and note size.

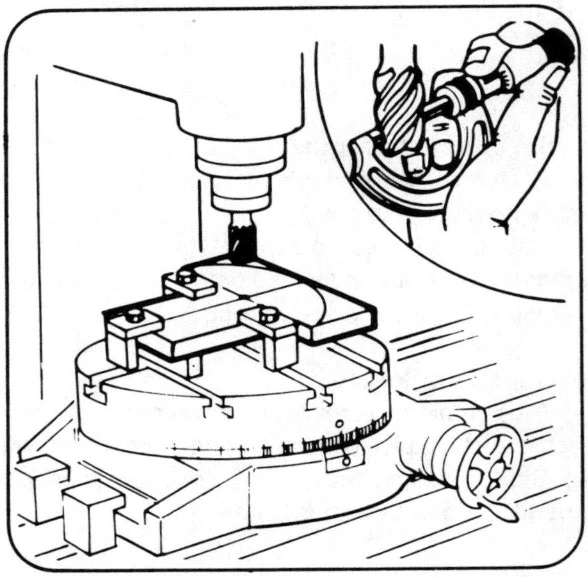

2 Position cutter to workpiece

(a) Move table slide required distance from center point, for finished size.

Example:

$$\frac{\text{diameter of cutter}}{2} + \begin{array}{l}\text{distance of radius}\\ \text{from center point}\end{array}$$

(b) Set table collar at zero and mark direction of movement.

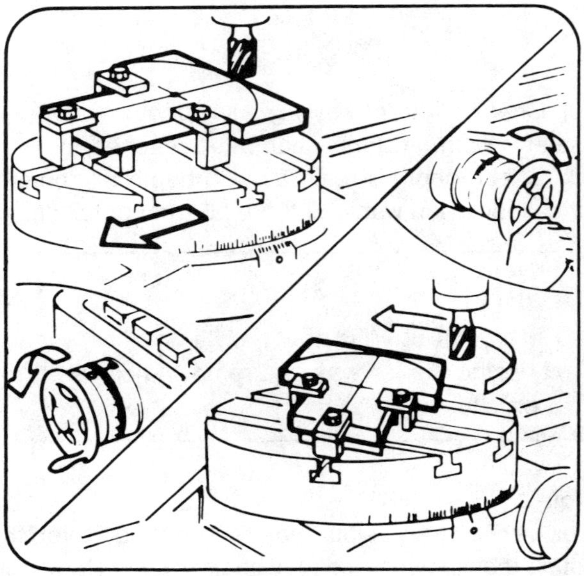

3 Finish face

(a) Move table slide to clear cutter from workpiece, and hand feed rotary table around to clear cutter from radius.

(b) Remove material in even cuts, using plenty of coolant.

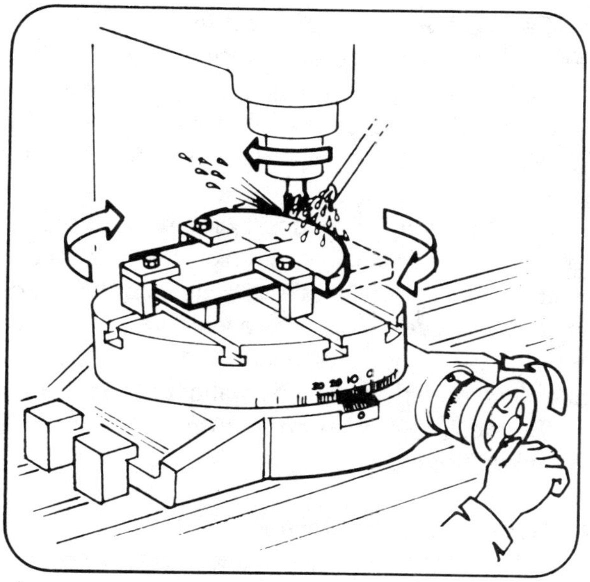

(c) Finish with fine cut, ensuring that final table setting is in direction marked.

Milling irregular shapes on the vertical mill

Irregular shaped workpieces, such as templates, which have curved and flat surfaces, and internal and external radii, are first machined to the appropriate rectangular size. The profile is then marked on the workpiece and finish machined on the rotary table. Centers of the radii are marked with a center and mounted in the machine spindle, the exact position being obtained by the machine table and saddle collars.

1 Machine workpiece to rectangular size

(a) Select suitable workholding method, and machine workpiece to appropriate rectangular size.

(b) Ensure that sides are square to each other.

2 Lay out workpiece

(a) Apply layout ink to workpiece.

(b) Mount workpiece on workholding device, and align reference face.

(c) Set to reference face, adjust table collars to radii centers, and mark with center mounted in machine spindle. Use reference marks for radii where applicable.

(d) Lay out curved and angular faces to complete profile.

3 Set workpiece for angular faces

(a) Mount rotary table on machine.
(b) Rotate table to zero scale on circumference. Zero handwheel collar and lock in position.
(c) Mount workpiece on suitable parallels, clear of table.
(d) Position workpiece approximately centered to rotary table, and visually align reference edge square or parallel. Tap workpiece to align reference face, using dial indicator to check alignment.
(e) Tighten clamp bolts, ensuring that complete angular face may be machined without hitting clamps.
(f) Recheck reference edge, to ensure that workpiece has not moved.

4 Machine angular faces

(a) Rotate rotary table to required angle, using scale; final setting by handwheel index. Lock.
(b) Position face to be machined to rear of machine, if possible.
(c) Adjust machine table to cover angular face with cutter.
(d) Remove material in even cuts until layout line is split.
(e) Calculate required movement of rotary table for subsequent angles, and machine in same manner.

Note:
Where more than one angular face must be machined, clamps may have to be moved. Always ensure that one clamp is moved at a time. Other clamps should be kept tight to ensure that the workpiece does not move.

Safety

Be careful of your hands when clamping the workpiece. Always use the correct size wrenches.

5 Set and machine radii and curved faces

Note:
It may be more convenient, especially with large workpieces, to machine angular faces before mounting on the rotary table. Small workpieces may be set in the correct central position for the radii before milling angles. Ensure that the rotary table is kept centered on the machine spindle.

Indexing heads

Indexing heads are used for a wide variety of machining operations where angular positioning of the workpiece is required. Examples include milling a square or hexagon on a shaft and drilling equally spaced holes on a given pitch circle diameter.

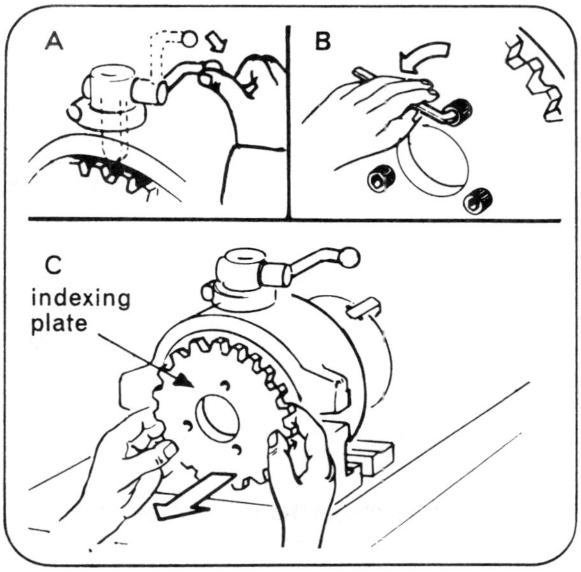

Changing the indexing plate

1 Remove indexing plate

(a) Tilt head over and lie it on its side with plunger lever uppermost.
(b) Rotate lever to disengage plunger from slot (A).
(c) Remove plate locking screws (B).
(d) Pull plate evenly at opposite sides to disengage from bore (C).

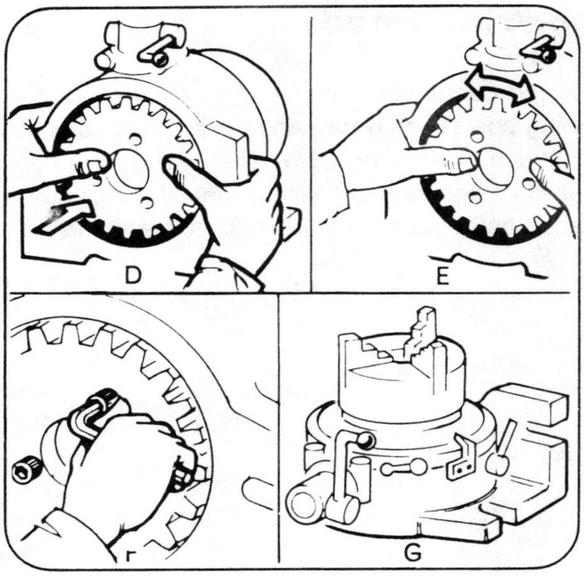

2 Position plate

(a) Select plate with appropriate number of slots. Number of slots must be equally divisible by divisions required.

(b) Wipe mating parts of plate and head clean and position plate on locator (D and E).

(c) Insert retaining screws in tapped holes and tighten securely (F).

(d) Wipe bottom face of head and table clean and tilt head up to set upright on table (G).

> **Safety**
>
> Take care not to trap your hands when tilting the head over.

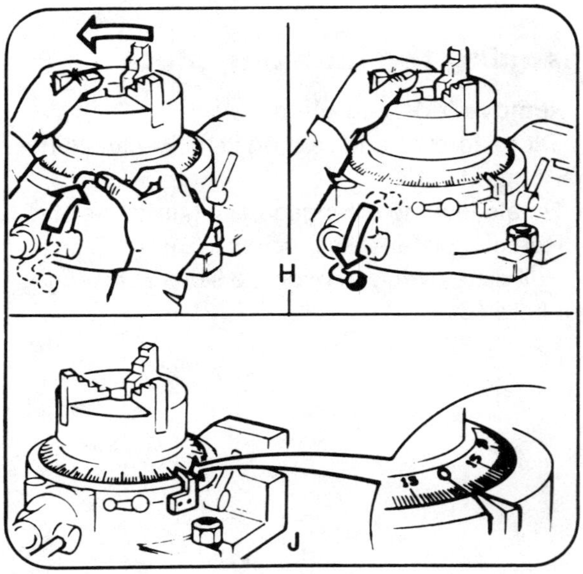

3 Set index

(a) Engage plunger in slot of indexing plate (H).

(b) Set pointer on graduated scale to zero (J).

Setting an indexing head on the table

(a) Place the head in the required position on the machine table.

(b) Align the slots of the table and the head. Bolt in vertical or horizontal position, as shown.

Safety

Take care not to drop the head. Keep your hands clear of the cutters.

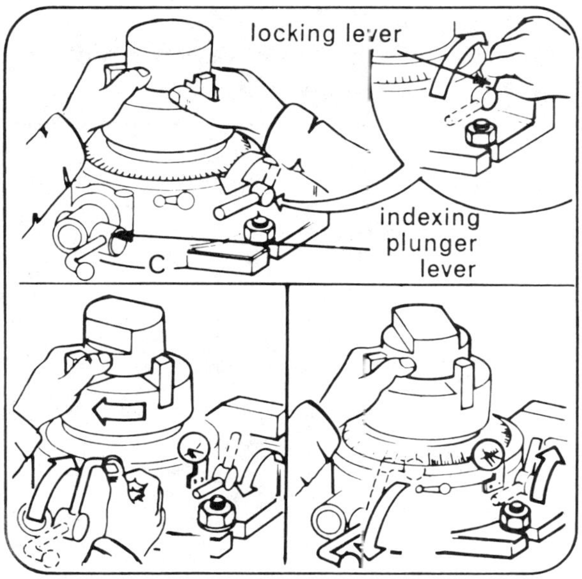

Indexing the chuck

(a) Make sure that correct indexing plate is installed.
(b) Free chuck by loosening locking lever, engage indexing plunger. Zero scale.
(c) Tighten chuck locking lever and take cut.
(d) Unlock chuck, disengage plunger, and rotate chuck until required division on graduated scale is approaching pointer on base.
(e) Release plunger lever and continue to rotate chuck until plunger engages in slot.
(f) Repeat steps (c), (d), and (e) until all faces have been machined.

Locating and clamping work in a chuck position workpiece

(a) Position and tighten workpiece in chuck with minimum amount of material extending.
(b) Protect machined surfaces.
(c) Work from reference if several components are to be machined.

Dividing heads

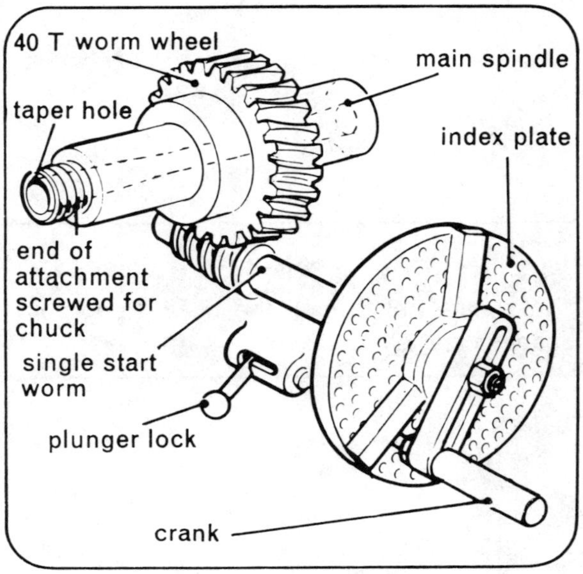

Dividing heads are used for milling slots, grooves, splines, teeth, and so on, which are equally spaced around the circumference of a workpiece. The divisions are obtained by index plates, of which there are usually three with the head. Many are reversible, with circles of holes on each side. The heads may also be geared to the machine lead screw, for milling spirals, cams, and so on. The workpiece may be held between centers or in one end in the chuck, supported by the tailstock center, or on an arbor fitted into the tapered hole in the spindle.

Slinging and lifting the dividing head

1 Set on machine table
(a) Clean table and base of head.

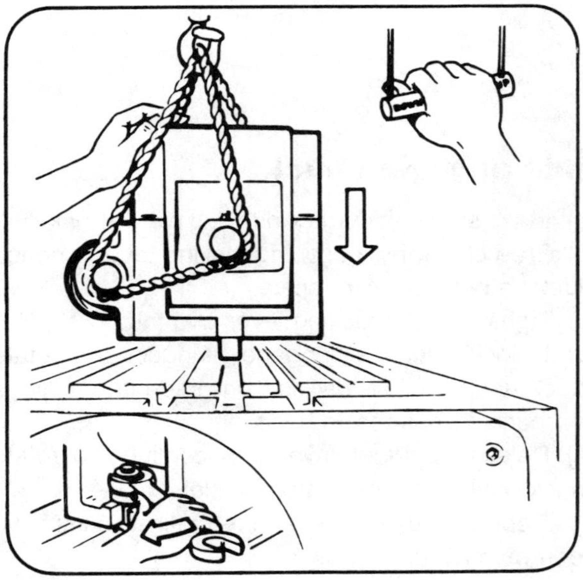

(b) Lower head to table, gently engaging keys in table slot.
(c) Tighten securing bolts.

> **Safety**
>
> Keep your hands clear of the bottom of the dividing head when lowering.

Setting the dividing head horizontal

1 Prepare head

(a) Zero index scale.

(b) Insert tapered shank setting bar in spindle, making sure that both are clean.

2 Check and correct setting

(a) Take head out of gear. Engage dial indicator as shown in diagram. Manually spin head to ensure that bar is running true. If not, remove bar and ensure that tapers on bar and in head are clean.

(b) Take reading with dial indicator as near spindle nose as possible.

(c) Repeat process at opposite end of bar. Note any variation.
(d) Remove stylus and slightly loosen locking nuts at back of head. Tap lightly with soft hammer in required direction, and check again.
(e) Repeat until indicator reading is same at each end of bar. Tighten locking nuts.

Setting the head to a given angle

1 Engage chuck

(a) Wipe threads and flange on spindle nose clean.

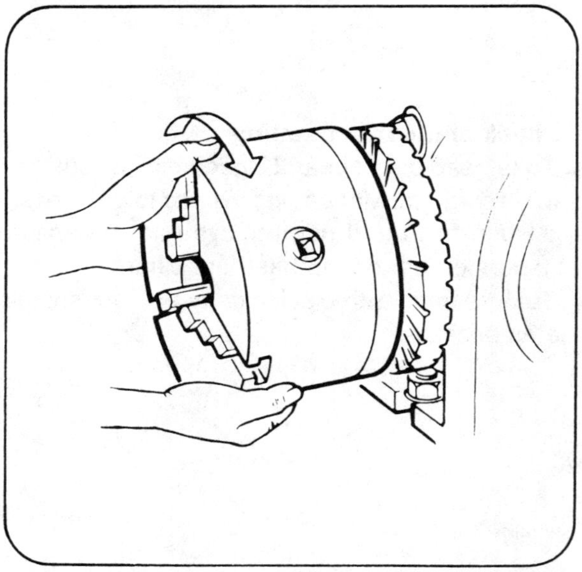

(b) Clean threads of chuck and screw tight onto spindle.

2 Set angle

(a) Loosen head locking nuts, and swivel head to the angle required on index scale.
(b) Finger tighten locking nuts, to hold head in position.

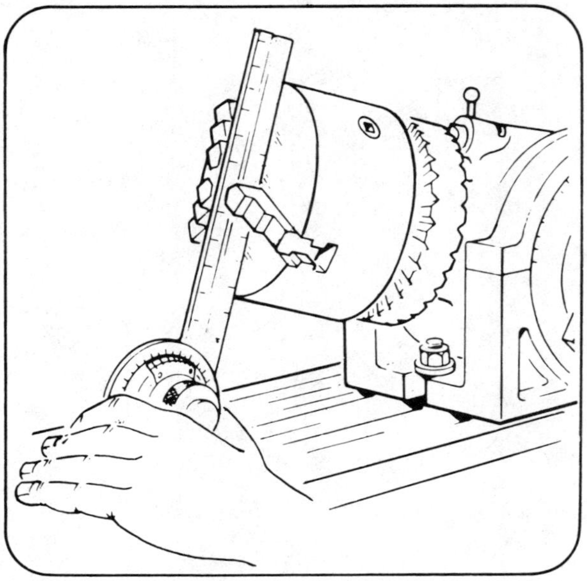

3 Check setting

(a) Set protractor to required angle. Example: if head is to be set at 30°, protractor will be set at 60°. Be sure that protractor base, chuck face, and table are clean.
(b) Tap chuck in required direction with soft hammer to achieve setting.
(c) Recheck.
(d) Tighten locking nuts securely.

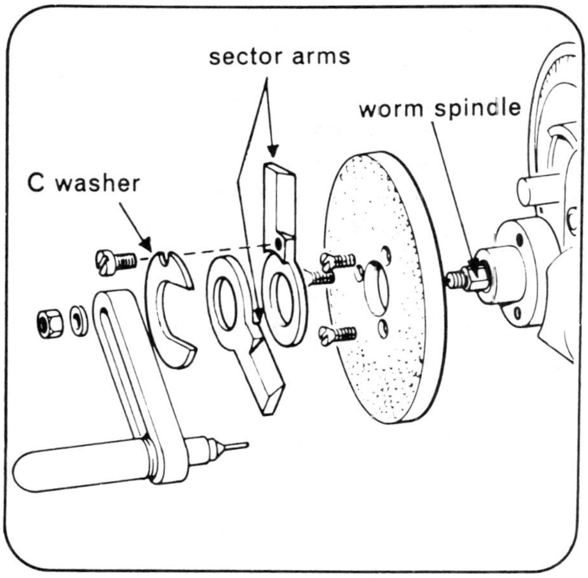

Changing the plate

1 Remove plate

(a) Remove nut and washer from end of worm spindle and slide off crank handle.
(b) Loosen locking screw in sector arms; remove C washer and sector.
(c) Remove the three screws from center of plate and release locking plunger at rear.
(d) Draw plate from worm spindle.

2 Position plate

(a) Select plate with required hole circle.
(b) Clean worm spindle and bore in plate.

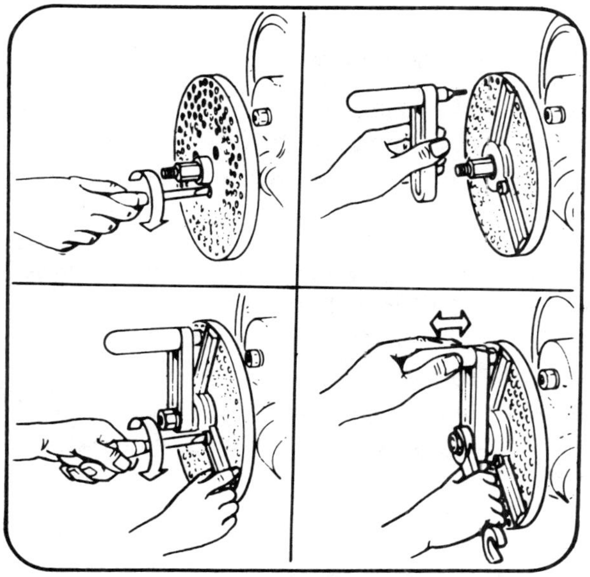

(c) Insert plate on worm spindle. Insert screws and tighten.
(d) Clean sector arms and C washer, and replace with locking screw.
(e) Replace crank, ensuring that index pin engages freely in required hole circle.
(f) Replace locking plunger at rear of plate.
(g) Replace washer and nut to end of worm spindle and tighten.
(h) Set sector arms to required number of holes and tighten. Check setting by counting holes between sectors after tightening.

Indexing the dividing head

There are three types of indexing:

1 Plain indexing with front plate (rapid indexing).

2 Plain indexing.

3 Differential indexing. (Consult a general text for differential procedures.)

Method 1 Plain indexing with front plate

(a) Take head out of gear. (Gear lever is at far side of head; not shown in diagram.)
(b) Clean slots in index plate.

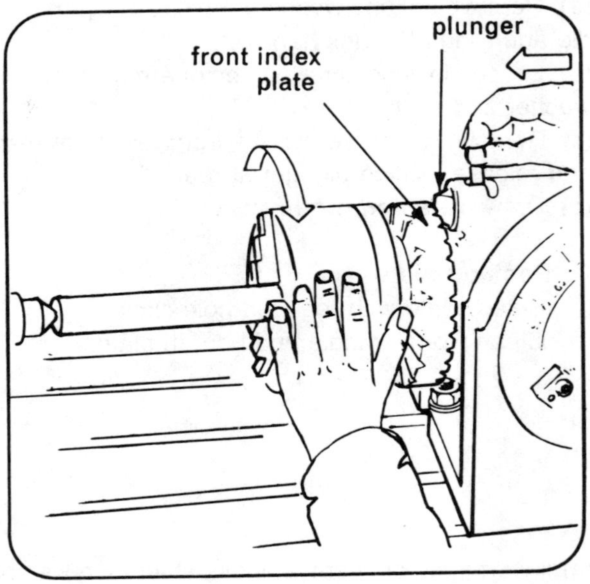

Note:
The diagram shows this index plate, which is different from the hole index plate previously referred to.
(c) Manually spin head and insert plunger in required slot.
(d) Lock head and take cut.
(e) Clear cutter from work. Unlock head, pull back plunger, and rotate chuck manually to required slot.
(f) Insert plunger, lock head, and take cut.

Note:
The front plate usually contains 24 slots and can be used to produce squares, hexagons, and other shapes. It is a much faster method than using the hole index plate.

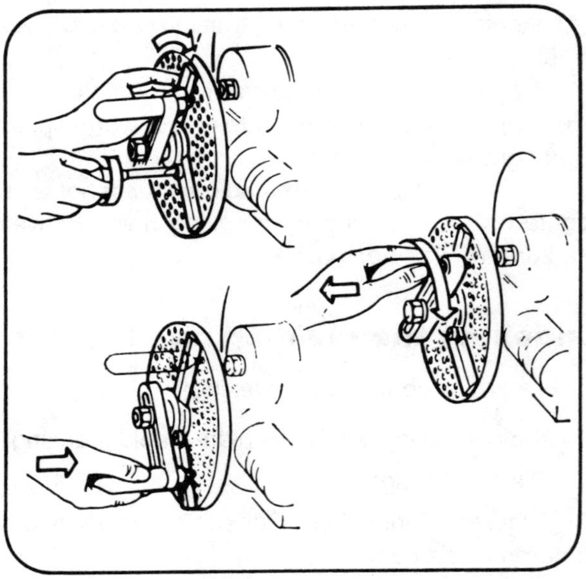

Method 2 Plain indexing

(a) Select suitable plate.
(b) Set sector arms to embrace number of holes required. Tighten locking screw.

Note:
The hole in which the index pin rests should NOT BE counted when setting the sectors.
(c) Turn crank clockwise required number of turns.
(d) Lock spindle and take cut.
(e) Move the sector around (clockwise) until other arm is touching plunger. It will then be ready for next indexing.

Note:
If the correct hole is overshot, the crank should be turned back at least half a turn and brought forward again to the correct hole. This will take up any backlash in the head worm.

Setting the tailstock horizontal

1 Set tailstock
(a) Clean table and base of tailstock.
(b) Position tailstock on machine table with keys in appropriate table slots.
(c) Insert taper shank setting bar in head and set head horizontal.
(d) Push up tailstock to within ½ in. (12 mm) of setting bar and tighten bolts.

(e) Crank up tailstock center to lightly engage in setting bar, visually checking that good entrance of center is obtained.

2 Check tailstock setting

(a) Take head out of gear and manually spin head to ensure good seating of tailstock center.

(b) Engage head in gear and check horizontal position of setting bar.

(c) Loosen locking nuts of tailstock. Gently adjust center (up and down) to bring dial indicator reading identical at each end of setting bar.

(d) Tighten locking nuts and recheck.

Setting the tailstock at an angle

1 Set angle

(a) Set head at required angle using head index graduated scale.

(b) Insert workpiece in head chuck. Loosen locking nuts on tailstock and adjust tailstock manually to an approximate angle and position to engage center in workpiece. Tighten locking nuts.

(c) Check angular setting with protractor. Adjust as necessary.

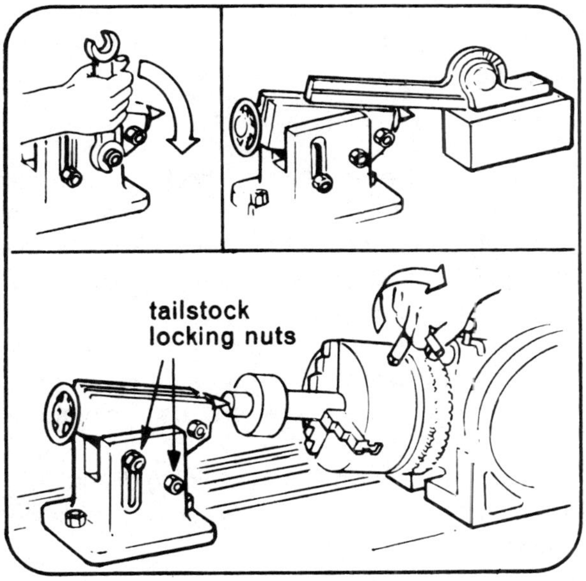

2 Install centers

(a) Using longitudinal and handwheel movements, engage tailstock center in workpiece.

(b) Ensure that tailstock center is seating equally around center in workpiece.

(c) Tighten table bolts.

Holding cylindrical workpieces between centers

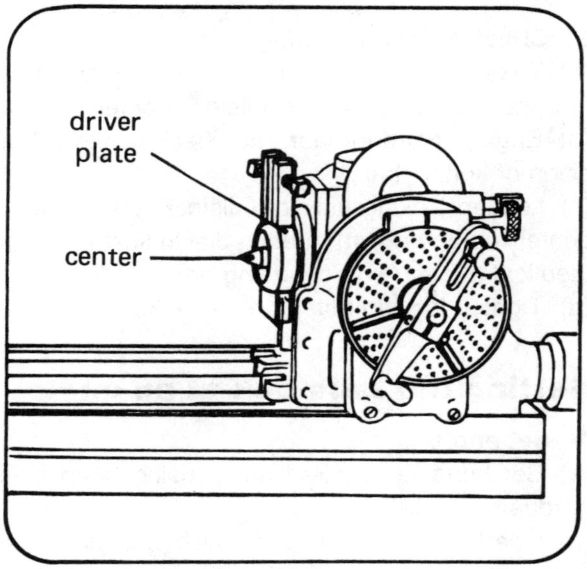

1 Mount dividing head horizontally

(a) Clean and locate center in dividing head spindle.

(b) Assemble driver plate, adjusting dog to eliminate backlash.

> **Safety**
>
> Keep your hands clear of the dividing head when lowering.

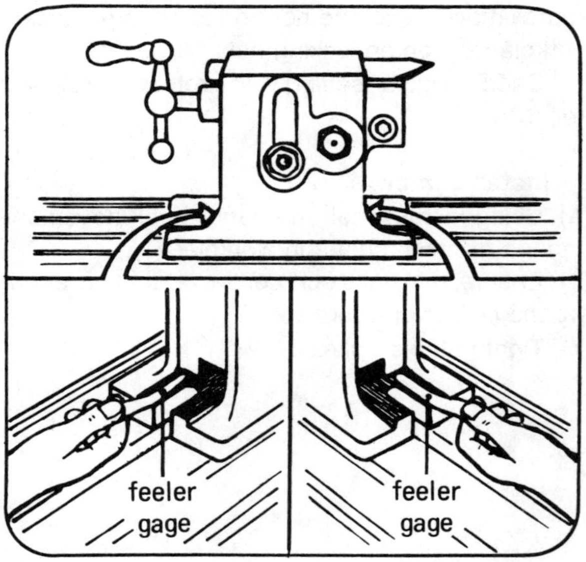

2 Mount tailstock

(a) Set to approximate position for length of workpiece.

(b) Check that center is in line and parallel with dividing head center.

Note:
This may be achieved either by checking with a feeler gage that the locating pads at the base of the movable section are in contact with the pads on the main casting of the tailstock, or by locating the taper pins provided for this purpose.

3 Mount workpiece

(a) Assemble carrier on shaft end, using a protective pad.

(b) Adjust set screws firmly, avoiding distortion.

(c) Adjust tailstock center firmly into workpiece center hole.

(d) Lock tailstock center.

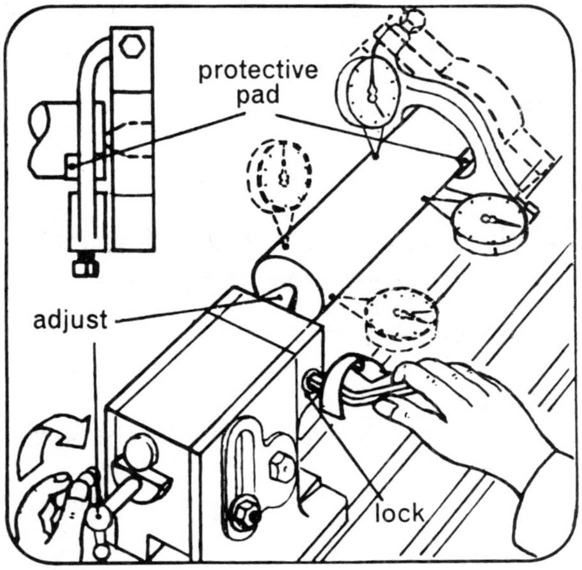

(e) Check each end in two axes, using dial indicator.

Holding cylindrical workpieces with chuck and centers

1 Mount dividing head with chuck horizontal

2A Positioning workpiece

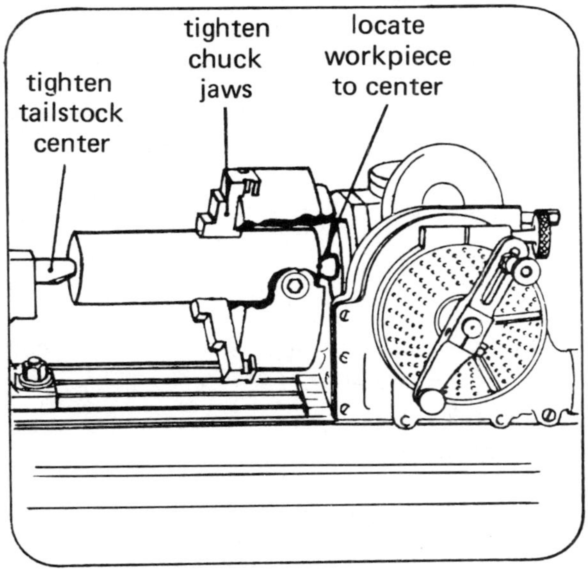

(a) Engage a center in dividing head spindle.
(b) Place workpiece in chuck, located lengthwise up to center.
(c) Tighten tailstock center.
(d) Tighten chuck jaws.

2B Positioning workpiece that is too short to reach center

(a) Cut out slots in substantial disk, to clear chuck jaws.
(b) Position disk at front face of chuck.

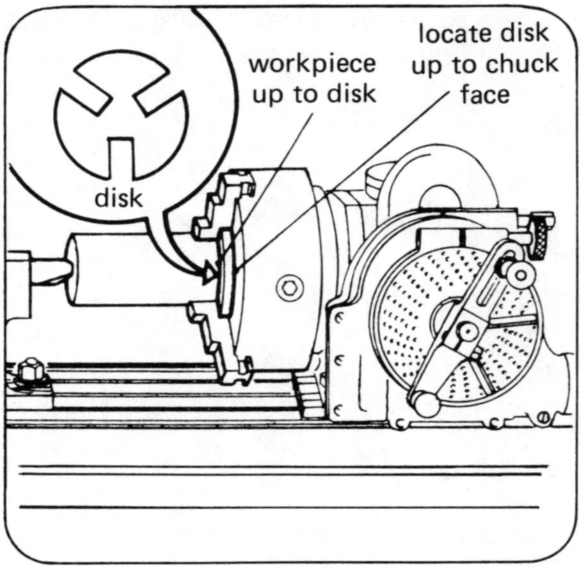

(c) Place workpiece in chuck firmly against disk face.
(d) Lightly tighten chuck jaws.
(e) Tighten tailstock center.
(f) Firmly tighten chuck jaws.

Setting tapered workpieces in collet and centers

1 Mount dividing head

2 Set angle

(a) Clean and engage taper shank test bar in head spindle.

(b) Loosen head locking nuts and, using index scale, swivel head to desired angle.
(c) Lightly tighten locking nuts, to hold head in position.

3 Check angle

(a) Calculate difference in height over a predetermined length and check, using dial indicator.

(b) Remove stylus and tap head face with soft hammer to correct any error.
(c) Recheck.
(d) Firmly tighten locking nuts.

4 Mount workpiece

(a) Remove test bar.
(b) Engage collet chuck with desired size collet.
(c) Mount workpiece.

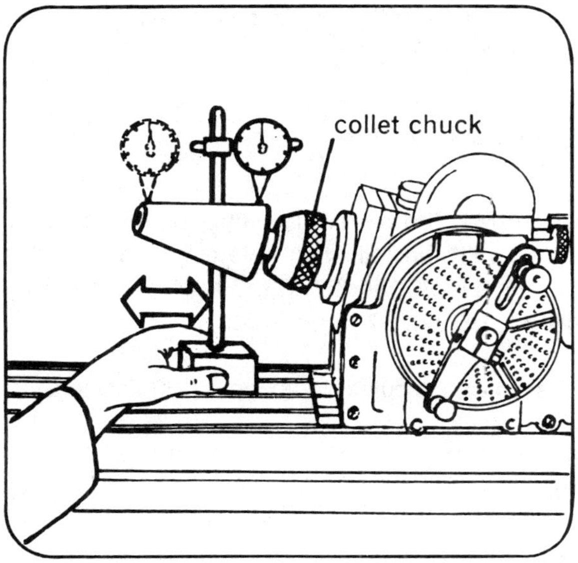

(d) Check, using dial indicator.

5 Set tailstock

(a) Release locking screw and slot nut.

(b) Rotate gear screw to operate rack.

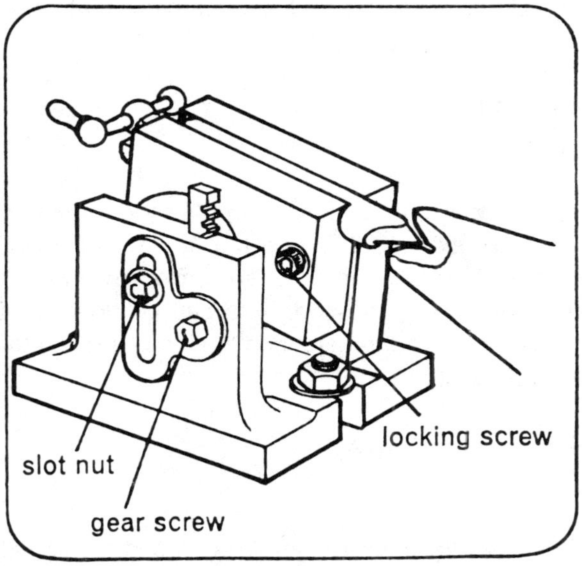

(c) Tilt to desired angle, and position at workpiece end visually.

(d) Tighten slot nut and register tailstock center in workpiece.

(e) Check, using dial indicator.

Note:

If the tailpiece center has not influenced the workpiece position, tighten the locking screw. If there has been any error, crank the center of the workpiece, tap the tailstock with a soft hammer, and recheck.

Calculations for indexing and dividing

Direct indexing

Direct indexing is the simplest method for dividing a circle into a required number of equally spaced divisions. The number of divisions that can be obtained by the direct method is limited by the number of holes in the index plate or in the spindle nose. Direct index head plates contain 24-, 30-, and 36-hole circles. The following formula may be used to make the index calculation:

$$\frac{\text{number of holes in the indexing circle}}{\text{required number of divisions}} = \frac{\text{holes}}{\text{division required for indexing}}$$

The result of this calculation must be a **whole num-**

ber because obviously only full holes are available into which the index may be engaged.

Examples: The available hole circles for direct indexing may be evenly divided in the following manner.

24 Divided by	30 Divided by	36 Divided by
2 = 12 holes/division	2 = 15 holes/division	2 = 18 holes/division
3 = 8	3 = 10	3 = 12
4 = 6	5 = 6	4 = 9
6 = 4		
8 = 3	6 = 5	6 = 6
12 = 2	10 = 3	9 = 4
24 = 1	15 = 2	12 = 3
	30 = 1	18 = 2
		36 = 1

Calculate direct indexing for 6 divisions:

$$\frac{24}{6} = 4 \text{ holes per division}$$

If you were using a 24-hole circle, the spindle would have to be rotated the partial turn corresponding to 4 holes. The 30- and 36-hole circles could also be used for 6 divisions. However, only the 30-hole circle could be used for 5 divisions and only the 36-hole circle for 9 divisions.

When direct indexing, count the required number of new holes to the next division. Do not count the hole in which the index pin is engaged. After engaging the pin, always lock the spindle to increase the rigidity of the setup. To avoid an indexing error, you may mark the hole circle with a felt pen or chalk.

Direct indexing may be done on a plain index head by disengaging the worm from the spindle so that the spindle can be turned by hand. The hole circles in the spindle nose are used for the indexing circles.

Simple indexing

Simple indexing, also known as plain indexing, makes use of the gear ratio between crank and spindle. On most common index heads this ratio is **40:1. That is, 40 turns of the index crank are required to rotate the spindle one revolution.**

Required numbers of circle divisions involving less than one full turn of the spindle now become fractions of 40, and the indexing calculation is made by applying the formula:

$$\text{index crank turns} = \frac{40}{N}$$

where N is the required number of divisions. Index plates with the following hole circles are available: 24, 25, 28, 30, 34, 37, 38, 39, 41, 42, 43, 46, 47, 49, 51, 53, 54, 57, 58, 59, 62, 66.

Example: Calculate indexing for 20 divisions.

$$\frac{40}{N} = \frac{40}{20} = 2 \text{ full turns of the index crank}$$

In this case any index plate may be used, since no partial turn is required. However, to calculate indexing for 52 divisions:

$$\frac{40}{N} = \frac{40}{52}$$

This means that less than one full turn is required for each division, $^{40}/_{52}$ of a turn to be exact. A 52-hole circle is not available, so a **proportional fraction** must be generated. This can be done by the following procedure. Reduce the index fraction to its lowest terms:

$$\frac{40}{52} = \frac{10}{13}$$

Take the **denominator** of the reduced fraction, 13, and determine into which of the available hole circles it can be **evenly divided.** In this case, 13 can be divided into the available 39-hole circle exactly 3 times. Use 3 as a **multiplier** to generate the proportional fraction required.

$$\frac{10 \times 3}{13 \times 3} = \frac{30}{39}$$

Therefore, **30 holes on a 39-hole circle** is the correct indexing for 52 divisions.

Example: Calculate indexing for 27 divisions.

$$\frac{40}{N} = \frac{40}{27} = 1\,\frac{13}{27}$$

or 1 full turn plus $^{13}/_{27}$ fractional turn. If a 27-hole circle were available, 13 holes of the 27 would give the fractional turn required. However, a 27-hole circle is not available, so a proportional fraction must be generated containing a denominator corresponding to an available hole circle.

The fraction $^{13}/_{27}$ is already at lowest terms, so take the denominator and determine which of the available hole circles it will divide into evenly. When divided into the 54 hole circle, 27 goes exactly 2 times. This result becomes the multiplier used to generate the proportional fraction.

$$\frac{13 \times 2}{27 \times 2} = \frac{26}{54}$$

Therefore, **26 holes on the 54-hole circle** will give the correct partial turn fraction. Indexing will then be one full turn plus 26 of the 54 holes for each of the required divisions.

Example: Calculate indexing for 35 divisions.

$$\frac{40}{N} = \frac{40}{35} = 1\,\frac{5}{35}$$

or 1 full turn plus $^{5}/_{35}$ fractional turn. Reducing $^{5}/_{35}$ to lowest terms yields $^{1}/_{7}$. The lowest term denominator 7 divides evenly into:

$$\left.\begin{array}{l}\text{28-hole circle (4 times)}\\ \text{42-hole circle (6 times)}\\ \text{49-hole circle (7 times)}\end{array}\right\}\ \text{multipliers}$$

Therefore, any of these indexing circles may be used by generating the proportional fractions in the following manner:

$$\frac{1 \times 4}{7 \times 4} = \frac{4}{28}\ \text{(4 holes on 28-hole circle)}$$
$$\frac{1 \times 6}{7 \times 6} = \frac{6}{42}\ \text{(6 holes on 42-hole circle)}$$
$$\frac{1 \times 7}{7 \times 7} = \frac{7}{49}\ \text{(7 holes on 49-hole circle)}$$

Example: Calculate indexing for 51 divisions.

$$\frac{40}{N} = \frac{40}{51}$$

Since a 51-hole circle is available, 40 holes on a 51-hole circle provides the correct indexing.

Using the index head

It would be very unhandy to have to count numbers of holes for fractional turns each time around. To facilitate matters, and to reduce possible errors in indexing, **sector arms** are used. The sector arms rotate about the index crank hub and are locked in place by a lock screw. The facing edges of the arms are beveled, and the number of holes required for fractional turns is established between the beveled edges. Remember that the number of holes required for a fractional turn will be **new holes.** The hole where the crank pin is engaged is not counted.

To operate the index head, unlock the spindle and pull back the crank pin plunger. Rotate the crank the required number of turns plus fractions. Set the sector arms to keep track of additional holes beyond one full turn of the crank. *Bring the crank pin around until it just drops into the required hole.* If you overshoot the position, back off well past the required hole and come up to it again. This will eliminate any backlash in the worm and wheel. **Always index in one direction only, and always lock the spindle before beginning the machining operation.**

After each indexing move, the sector arms must be shifted so that they will be properly positioned for the next move. For example, if you are indexing one full turn plus 11 holes, set the sector arms to include **11 new holes** between the beveled faces. Turn the index crank one full turn plus the 11 additional holes. Then, shift the sector arms around so that 11 new holes will be available for the next move.

Direct angular indexing

Since there are 360 degrees in a circle, each hole in any index plate circle can be equated to degrees and fractions by the formula

$$\text{degrees per hole} = \frac{360}{N_H}$$

where N_H **= number of holes in the index circle.** Transposing this formula to determine the number of holes corresponding to an angle gives

$$N_H = \frac{360}{\text{degrees in angle desired}}$$

Example: Calculate direct indexing for 15°.

$$N_H = \frac{360}{15} = 24$$

Therefore, a 24-hole circle would equate to 15° per hole.

To determine the number of index plate holes that define a specific angle, **divide the degrees per hole of a specific index circle into the angle.**

Example: Calculate indexing for 75° using a 24-hole index circle.

$$\frac{75°}{15°/\text{hole}} = 5 \text{ holes for } 75°$$

Simple angular indexing

On the simple index head making use of a 40–1 ratio, 40 turns of the index crank are required to turn the spindle 360° or one full revolution. Therefore, one turn of the index crank is equal to

$$\frac{360°}{40 \text{ turns}} = 9°/\text{turn}$$

Any index plate circle that is divisible by 9 can be used for angular indexing.

Example:

$$\frac{\text{27-hole circle}}{9} = 3 \text{ holes per degree}$$

(1 hole = ⅓° or 20 min)

$$\frac{\text{36-hole circle}}{9} = 4 \text{ holes per degree}$$

(1 hole = ¼° or 15 min)

$$\frac{\text{45-hole circle}}{9} = 5 \text{ holes per degree}$$

(1 hole = ⅕° or 12 min)

$$\frac{\text{54-hole circle}}{9} = 6 \text{ holes per degree}$$

(1 hole = ⅙° or 10 min)

The index crank turns required for an angle can be calculated by the formula:

$$\text{turns} = \frac{\text{degrees required}}{9°/\text{turn}}$$

Example: Calculate indexing for 37 degrees.

$$\frac{37}{9} = 4\frac{1}{9}$$

or 4 full turns plus 1/9 additional turn. The 1/9 partial turn is found by the same method described in the preceding section. Determine which of the available index circles the lowest term denominator 9 will divide into evenly.

$$\frac{\text{54-hole circle}}{9} = 6 \text{ holes}$$

Therefore, indexing is **4 full turns plus 6 holes** on a 54-hole circle.

Angular indexing in minutes

Since one full turn of the index crank in a 40:1 ratio head is equal to 9°, **it is also equal to 9 × 60 or 540 minutes.** Therefore, to index by minutes' apply the formula:

$$\text{crank turns} = \frac{\text{minutes required}}{540'/\text{turn}}$$

Examples: Calculate indexing for 8°, 50′. Convert angle to minutes:

$$8°\ 50' = 530'$$

Apply the formula:

$$\text{crank turns} = \frac{\text{minutes required}}{540'/\text{turn}} = \frac{530'}{540'}$$

Reduce the fraction to lowest terms:

$$\frac{530}{540} = \frac{53}{54}$$

Take the lowest common denominator 54, and determine the available hole circle that it will divide into evenly. Since a 54-hole circle is available, 53 holes on a 54-hole circle will index 8° 50′.

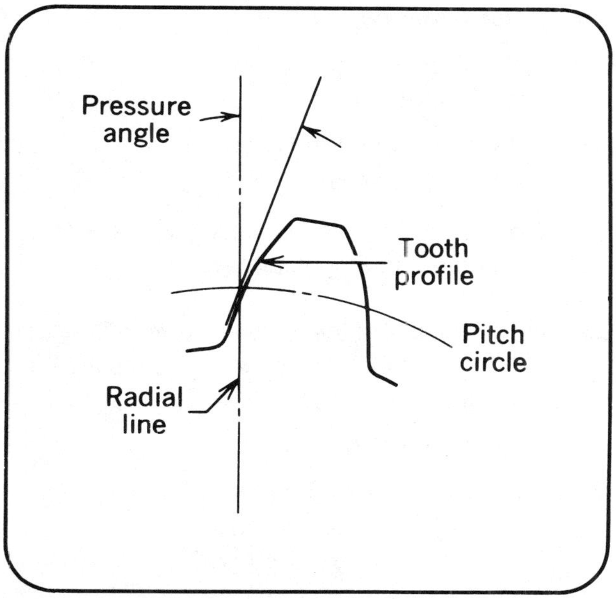

Spur gear cutting

Involute gear cutters

Spur gears are cut with **involute gear cutters.** The gear cutter set consists of eight cutters numbered 1 to 8, for typical pressure angles of 14½ or 20°. The eight gear cutters have eight **different** tooth forms for each diametral pitch, depending on the number of teeth on the gear to be cut. Each gear cutter can be used to cut a range of teeth (Table 1). However, the cutter will produce the **exact tooth form** only for the **lowest number of teeth in the range.** For example, a 5 cutter will cut 21 to 25 teeth, but will produce the exact form only for 21 teeth. Other numbers of teeth will be less accurate in form.

Half-step cutters are also available. These serve to cut more accurate tooth forms within a cutter's range (e.g., a 5½ for 19 and 20 teeth). Special cutters designed for specific numbers of teeth or for low numbers of teeth (6–11) are also used.

Gear cutters are identified and marked with the following information.

Cutter number

Diametral pitch

Number of teeth

Pressure angle

Whole depth of tooth

Table 1 Involute Cutters

Number of Cutter	Number of Teeth Cut	Number of Cutter	Number of Teeth Cut
1	135 to rack	5	21–25
2	55–134	6	17–20
3	35–54	7	14–16
4	26–34	8	12 and 13

Spur gear calculations

The following symbols are used to represent gear tooth terms in spur gear calculations.

P = **diametral pitch**

D = **pitch diameter**

D_o = **outside diameter**

N = **number of teeth in gear**

t = **tooth thickness, circular**

a = addendum

b = dedendum

c = clearance

C = center distance

h_k = working depth

h_t = whole depth

t_c = tooth thickness, chordal

a_c = addendum, choral

Table 2 gives the formulas to calculate the gear dimensions in the following examples.

Example 1: Determine the dimensions for a 30-tooth gear, 14½° pressure angle, and a 2.500 in. pitch diameter.

1 Number of teeth N = 30.

2 Pitch diameter D = 2.500 in.

3 $$\frac{\text{Diametral pitch}}{P} = \frac{N}{D} = \frac{30}{2.500} = 12$$

4 $$\frac{\text{Addendum}}{a} = \frac{1}{P} = \frac{1}{12} = .083 \text{ in.}$$

5 $$\frac{\text{Dedendum}}{b} = \frac{1.157}{P} = \frac{1.157}{12} = .096 \text{ in.}$$

6 Tooth thickness

$$t = \frac{1.5708}{P} = \frac{1.5708}{12} = .131 \text{ in.}$$

7 $$\frac{\text{Clearance}}{c} = \frac{.157}{P} = \frac{.157}{12} = .013 \text{ in.}$$

8 $$\frac{\text{Whole depth}}{h_t} = \frac{2.157}{P} = \frac{2.157}{12} = .179 \text{ in.}$$

9 $$\frac{\text{Working depth}}{h_k} = \frac{2}{P} = \frac{2}{12} = .166 \text{ in.}$$

10 Chordal tooth thickness

$$t_c = D \sin \frac{90^\circ}{N}$$

$$= 2.500 \times \sin \frac{90^\circ}{30}$$

$$= 2.500 \times \sin 3^\circ$$

$$= 2.5 \times .0523 = .1307 \text{ in.}$$

11 Chordal addendum

$$a_c = a + \frac{t^2}{4D}$$

$$= .083 + \frac{.131^2}{4 \times 2.5}$$

$$= .083 + \frac{.017}{10}$$

$$= .083 + .0017 = .0847 \text{ in.}$$

12 Outside diameter

$$D_o = \frac{N + 2}{P}$$

$$= \frac{30 + 2}{12}$$

$$= \frac{32}{12} = 2.6666 \text{ in.}$$

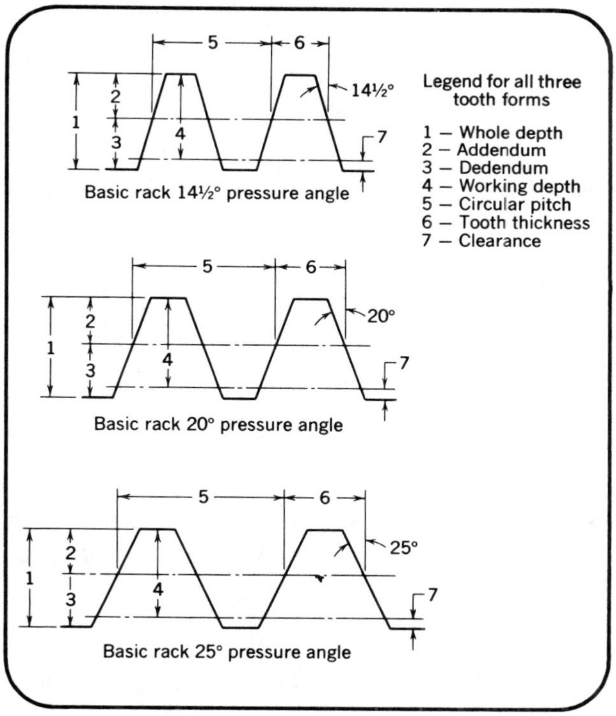

Table 2 Spur Gear Formulas

To Find	Spur Gear Formulas: – 14½° Pressure Angle	Spur Gear Formulas: 20 and 25° Pressure Angles
Addendum, a (in.)	$a = \frac{1.0}{P}$	$a = \frac{1.0}{P}$
Dedendum, b (in.)	$b = \frac{1.157}{P}$	$b = \frac{1.250}{P}$
Pitch diameter, D	$D = \frac{N}{P}$	$D = \frac{N}{P}$
Outside diameter, D_o	$D_o = \frac{N + 2}{P}$	$D_o = \frac{N + 2}{P}$
Number of teeth, N	$N = D \times P$	$N = D \times P$
Tooth thickness, t (in.)	$t = \frac{1.5708}{P}$	$t = \frac{1.5708}{P}$
Whole depth, h_t (in.)	$h_t = \frac{2.157}{P}$	$h_t = \frac{2.250}{P}$
Clearance, c (in.)	$c = \frac{.157}{P}$	$c = \frac{.250}{P}$
Center distance, C	$C = \frac{N_1 + N_2}{2 \times P}$	$C = \frac{N_1 + N_2}{2 \times P}$
Working depth, h_k	$h_k = \frac{2}{P}$	$h_k = \frac{2}{P}$
Chordal tooth thickness, t_c	$t_c = D \sin\left(\frac{90°}{N}\right)$	$t_c = D \sin\left(\frac{90°}{N}\right)$
Chordal addendum, a_c	$a_c = a + \frac{t^2}{4D}$	$a_c = a + \frac{t^2}{4D}$
Diametral pitch, P	$P = \frac{N}{D}$	$P = \frac{N}{D}$
Center distance, C	$C = \frac{D_1 + D_2}{2}$	$C = \frac{D_1 + D_2}{2}$

Example 2: Determine the gear dimensions for a 45-tooth gear, 8 diametral pitch, 20° pressure angle.

1 Number of teeth $N = 45$

2 Pitch diameter $D = N/P = 45/8 = 5.625$ in.

3 Diametral pitch $P = 8$

4 Addendum $a = 1/P = 1/8 = .125$ in.

5 Dedendum $b = 1.250/P = 1.250/8 = .1562$ in.

6 Tooth thickness

$$t = \frac{1.5708}{P} = \frac{1.5708}{8} = .1963 \text{ in.}$$

7 Clearance $c = .250/P = .250/8 = .031$ in.

8 Whole depth $h_t = 2.250/P = 2.250/8 = .281$ in.

9 Working depth $h_k = 2/P = 2/8 = .250$ in.

10 Chordal tooth thickness

$$
\begin{aligned}
t_c &= D \sin \frac{90^\circ}{N} \\
&= 5.625 \times \sin \frac{90^\circ}{45} \\
&= 5.625 \times \sin 2^\circ \\
&= 5.625 \times .0349 = .1963 \text{ in.}
\end{aligned}
$$

11 Chordal addendum

$$
\begin{aligned}
a_c &= a + \frac{t^2}{4D} \\
&= .125 \text{ in.} + \frac{.1963^2}{4 \times 5.625} \\
&= .125 \text{ in.} + .0017 \text{ in.} = .1267 \text{ in.}
\end{aligned}
$$

12 Outside diameter

$$
\begin{aligned}
D_o &= \frac{N + 2}{P} \\
&= \frac{45 + 2}{8} \\
&= \frac{47}{8} = 5.875 \text{ in.}
\end{aligned}
$$

Cutting a spur gear

Machine a gear blank to the correct outside diameter and mount on the arbor. Be sure to lubricate the bore of the blank before pressing.

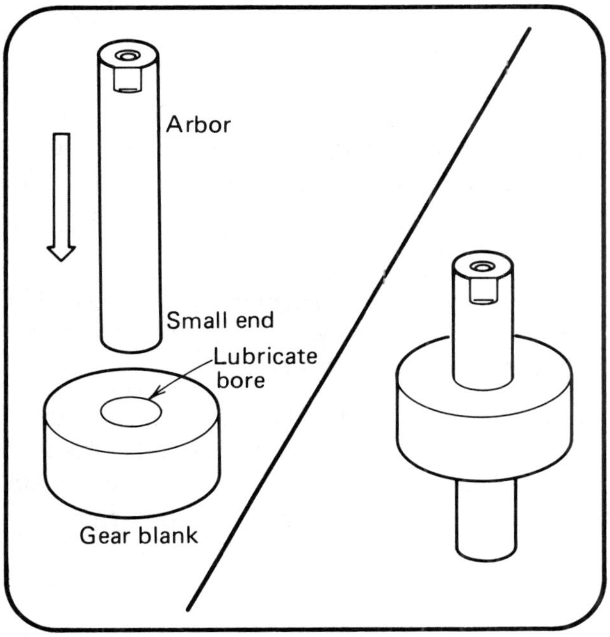

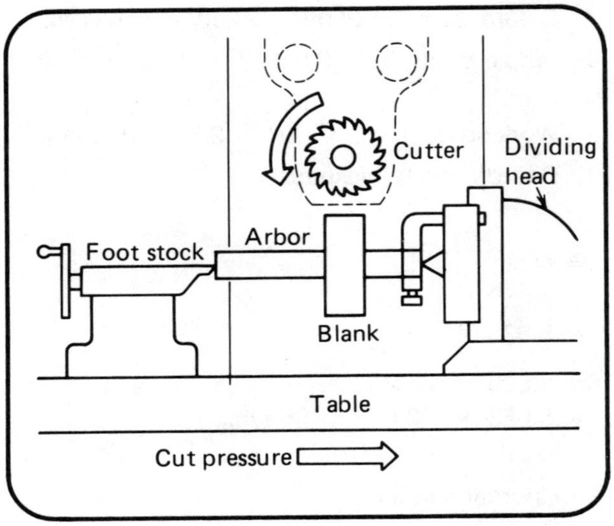

Install the arbor between the centers of the dividing head and tailstock. The dividing head must be aligned parallel to the machine table.

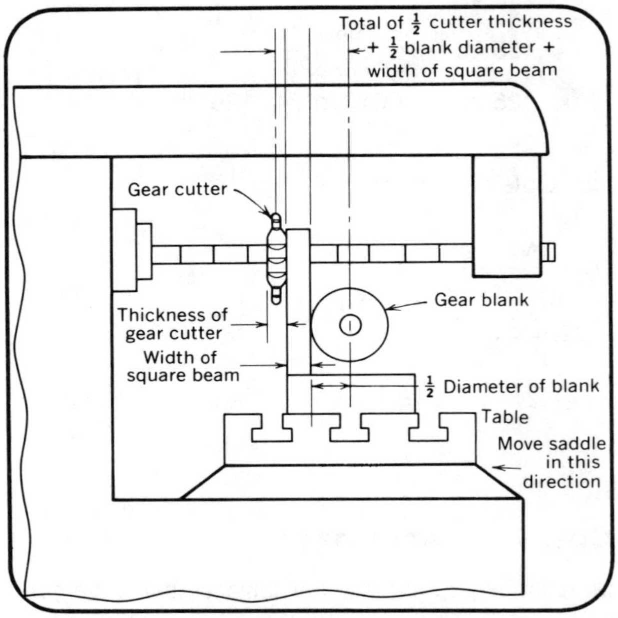

Mount the correct gear cutter on the machine arbor. Install the arbor support and tighten the arbor nut. Center the gear blank under the gear cutter.

Calculate the required indexing and install the correct plate. Determine the whole depth of the gear teeth required.

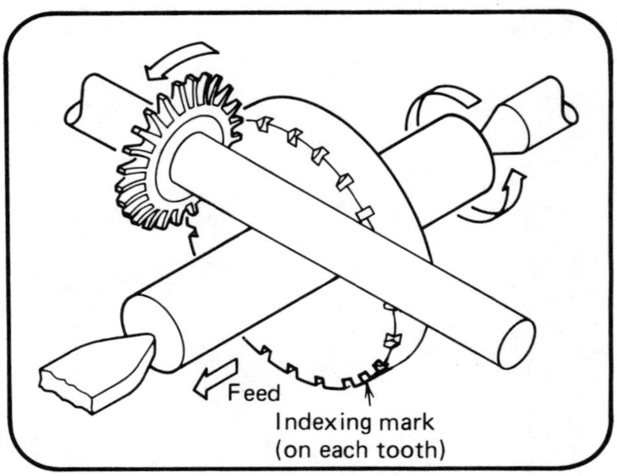

Index to each tooth position and touch the cutter to the corner of the blank to leave a small mark. When all tooth positions have been marked, check that indexing is correct.

Make roughing cuts on all teeth leaving about .040 in. (1 mm) for finishing. Be sure to direct cut pressure toward the dividing head spindle.

Raise knee to final depth minus a few thousandths. Mill two adjacent gear teeth and deburr. Use the gear tooth vernier to measure tooth thickness. Tooth thickness should be slightly oversized because full depth has not yet been set on the knee.

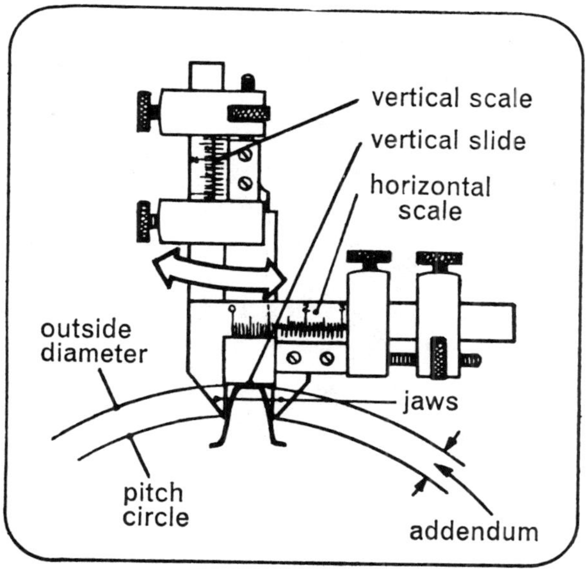

Set up gear tooth vernier

(a) Adjust horizontal vernier scale to set jaws of vernier to tooth thickness required.

(b) Set vertical vernier scale to addendum.

(c) Place verniers on tooth. Ensure that vertical slide rests on outside diameter.

(d) Move verniers back and forth. Close horizontal vernier if any movement; open horizontal vernier if clearance seen under vertical slide. Check measurement.

Raise the knee to the final depth setting and mill the two adjacent teeth. Check for correct tooth dimension and if correct, final mill the remaining teeth.

Review questions

1 Describe applications of the rotary table.

2 Describe and demonstrate alignment techniques for the rotary table.

3 Describe and demonstrate the use of the index head.

4 Describe and demonstrate the use of the dividing head.

5 How may a workpiece be held during a dividing operation?

6 Describe and demonstrate indexing and dividing head calculations.

7 What is angular indexing and how is it used?

8 How are involute gear cutters selected?

9 Describe in general terms spur gear cutting on the milling machine.

10 How may the dividing head be used to set work at an angle?

UNIT ELEVEN

Milling Faults and Their Correction

Explanation

To be able to attain certain quality standards, you should know faults that can, and do, occur, and how to correct them.

Because of the wide variety of work, and the many types of machine and pieces of associated equipment used, it is not feasible to give the cause of each fault. Only examples of common faults are shown.

The first column headed "Fault" gives the name and description of the fault.

The second column headed "Early Warning" describes warning signs of a fault developing and how it may be related to a particular cause.

The cause of the fault is given in the third column.

The last column gives preventive and corrective actions.

The emphasis is on prevention rather than correction, the checking of possible causes of faults, and knowing how they may be prevented.

Milling Faults and Their Correction

FAULT	EARLY WARNING	CAUSE	PREVENTION AND/OR CORRECTION
Chatter Lateral grooves and ridges, giving mottled appearance.	Hear cutting sound change to vibrating noise. This is common to all causes.	Wrong cutter selected.	Use coarser tooth cutter.
		Speed too fast.	Decrease spindle speed.
		Feed too fine.	Increase rate of feed.
		Cutter worn or damaged.	Check cutter for wear or damage.

Milling Faults and Their Correction (*Continued*)

FAULT	EARLY WARNING	CAUSE	PREVENTION AND/OR CORRECTION
Chatter (*cont.*)		Cutter incorrectly ground.	Check that cutter is correctly ground.
	See cutter stalling.	Cutter loose on arbor.	Check that cutter is secure on arbor.
		Arbor loose in spindle.	Check that arbor is secure in spindle.
		Play in spindle bearings.	Check for play in spindle bearings.
	Feel vibration between fixed and movable parts of vertical head.	Quill and/or vertical head not locked.	Check that quill and/or vertical head is locked.
	Feel vibration between workpiece and workholding device.	Workpiece not securely clamped.	Check that workpiece is securely clamped.
	Feel workpiece vibrating.	Excessive overhang of workpiece.	Support excessive overhang of workpiece.
	Feel vibration between workholding device and machine table.	Workholding device loose.	Check that workholding device is securely clamped.
	Feel vibration between fixed and movable parts of workholding device.	Locking levers of workholding device loose.	Check that locking levers of workholding device are tightened.
	Feel easy movement of table when setting.	Play in table gibs.	Check for play in table gibs.
	Feel vibration between gibs and slides.	Locking levers loose.	Check that locking levers tight except on moving component.
Coarse Finish Torn grooves and ridges on face.	See ragged or discolored chips.	Worn cutter.	Check to see if cutter teeth are dull and/or discolored.
		Cutter teeth damaged.	Check cutter for chipped or broken teeth.
	See chips building up between cutter teeth.	Insufficient clearance on cutter teeth.	Check amount of clearance on cutter teeth.
	Hear intermittent cut.	High tooth in cutter.	Check for high tooth in cutter.
Cutter marks widely spaced.		Feed too coarse.	Decrease feed rate.
		Speed too slow.	Increase spindle speed.

FAULT	EARLY WARNING	CAUSE	PREVENTION AND/OR CORRECTION
Coarse Finish Cutter marks widely spaced (*cont.*)	See discolored chips.	Insufficient coolant.	Increase coolant flow.
	See jerky table movement.	Irregular feed rate.	Check that locking levers are not tightened. Check that gibs are not overtightened.
Taper Faces not parallel.	See cutter remove metal from one side or end only, when touching to face.	Chips or burrs on face of table.	Check that face of table is clean and free of burrs. Feel for rock of workholding device or workpiece when locating.
	It may also be possible to see an increasing or decreasing depth of cut if only a light cut is taken.	Chips or burrs on bottom face of workholding device.	Check that bottom face of workholding device is clean and free of burrs. Feel for rock when locating.
	Often no early warning signs are indicated, so check for taper before taking finish cut.	Chips or burrs on locating face(s) of workholding device.	Check that locating face(s) of workholding device is clean and free of burrs. Feel for rock when locating workpiece.
		Chips or burrs on locating face(s) of workpiece.	Check that locating face(s) of workpiece is clean and free of burrs. Feel for rock when locating.
		Chips or burrs on face(s) of parallel(s).	Check that face(s) of parallel(s) are clean and free of burrs. Feel for rock when locating.
		Workpiece not correctly seated.	Check seating with feeler gage. Feel for movement of parallel(s).
		Workholding device or workpiece incorrectly aligned.	Check setting of workholding device and/or workpiece in appropriate axes.
		Machine spindle not set vertical.	Check vertical setting of machine spindle.
		Parallels of different heights.	Check that equal size parallels used.
		Parallel(s) not parallel.	Check parallelism of parallel(s).
		Periphery of cutter not ground parallel.	Check parallelism of cutters if cutting on periphery.

Milling Faults and Their Correction (*Continued*)

FAULT	EARLY WARNING	CAUSE	PREVENTION AND/OR CORRECTION
Dimensional inaccuracies		Inaccurate checking of prefinished dimension.	Check that correct reading is obtained.
		Inaccurate calculation of amount to be removed.	Check accuracy of calculations.
		Inaccurate calculation of amount of table movement.	Check accuracy of calculations.
		Inaccurate amount of table movement on index.	Check index reading before and after moving table.
		Inaccurate setting of reference.	Check accuracy of setting of reference face, bore, or step.
		Wrong size cutter selected.	Check that size of cutter selected is correct.
		Machine spindle not set vertical.	Check vertical setting of machine spindle.
		Inaccurate touching of setting bar or cutter to workpiece.	Check touch of cutter or setting bar to workpiece. Check thickness of feeler(s) if used.
		Setting bar or cutter not running true.	Check true running of setting bar or cutter(s).
		Inaccurate spacing of cutters.	Check accuracy of distance between cutters if more than one is used.
	See index ring not rotating with table movement.	Index ring loose.	Check that index is tightened.
	Feel direction of free movement of lead screw.	Backlash wrong way.	Check direction of backlash and accommodate it when moving table. Move table toward cutter thrust when positioning.
		Inaccuracy of lead screw.	Check accuracy of index over a given distance.
		Inaccurate measuring instruments.	Check accuracy of measuring instrument.

FAULT	EARLY WARNING	CAUSE	PREVENTION AND/OR CORRECTION
Dimensional inaccuracies (*cont.*)		Movement of cutter while cutting.	Check that cutter is properly located and secure.
Faces not square Light seen between edges of square and workpiece when checked.		All causes of taper.	
		Universal table incorrectly aligned.	Check alignment of table.
		Inaccurate setting or measuring equipment.	Check accuracy of square, if used, for setting or checking.
Faces not flat		Workpiece distorted through clamping.	Check that workpiece supported under clamps. Release clamping pressure on workpiece and tighten only lightly before taking finishing cut.
		Internal stresses of workpiece released by machining.	Release clamping pressure on workpiece and tighten only lightly before taking finish cut. If possible, rough opposite faces of workpiece before finishing.
Angles incorrect		Wrong cutter selected.	Check angle of cutter selected.
		All causes of taper.	
		Protractor, if used, incorrectly set.	Check accuracy of setting of protractor.
		Sine bar, if used, incorrectly set.	Check calculation of blocks required. Check that correct blocks selected.
		Chips or burrs on setting instruments or on locating faces.	Check that setting instruments clean and free of burrs. Check that locating faces clean and free of burrs.
		Workpiece or workholding device incorrectly aligned to setting instruments.	Check setting of workpiece or workholding device and correctly align.
		Indexed workholding device inaccurately set.	Check that workholding device correctly zeroed when aligning to appropriate table movement.

Milling Faults and Their Correction (*Continued*)

FAULT	EARLY WARNING	CAUSE	PREVENTION AND/OR CORRECTION
Angles incorrect (*cont.*)		Indexed workholding device inaccurately set to index.	Check that workholding device is correctly indexed.
		Setting instrument incorrectly positioned.	Check that edge of setting instrument is square or parallel to edge of face being aligned.
		Adjustable parts of machine incorrectly set.	Check angular setting of universal table or vertical head if used to obtain angle.
Drilled hole undersize		Wrong drill selected.	Check size of drill.
	Hear drill whistling. Feel drill binding.	Worn drill.	Check for wear on periphery of drill.
Drilled hole oversize		Wrong drill selected.	Check size of drill.
	See drill wobble at start of cut.	Point of drill not centered.	Check that drill point is ground centered.
		Cutting edges unequal length.	Check height of cutting edges on periphery.
		Point or cutting edge of drill chipped.	Check drill for chipped point or cutting edges.
	See drill is not running concentric.	Bent drill.	Check that drill is not bent.
		Chips or burrs on drill shank.	Check that drill shank is clean and free of burrs.
		Chips or burrs on sleeves.	Check that sleeves, if used, are clean and free of burrs.
		Chips or burrs on arbor.	Check that arbor is clean and free of burrs.
		Chips or burrs in spindle bore.	Check that spindle bore is clean and free of burrs.
		Chips or burrs on drill chuck.	Check drill chuck, if used, for chips or burrs on jaws and shank.
	See discolored chips.	Insufficient coolant.	Increase coolant flow.

FAULT	EARLY WARNING	CAUSE	PREVENTION AND/OR CORRECTION
Drilled hole not square to face	See causes of faces not square. See causes of drilled hole undersize and oversize.		

Glossary

Alloy Two or more metals chemically combined

Blind hole A hole that does not pass through a workpiece

Blow hole In a casting, a hole caused by a gas bubble; porosity

Chamfer An angle machined on the corner of a workpiece, or at the entrance to a hole; a corner break

Chatter Noise and vibration caused by incorrect workholding, incorrect feeds and speeds, dull cutters, and other factors

Clinometer A measuring tool for determining angles

Collar The graduated dial on the table saddle and knee threads, used to measure movement of the components

Concave A surface curving below another surface or inward

Convex A surface curving above another surface or outward

Cutting speed The rate at which a point on the cutting tool passes the work in a given amount of time, usually measured in surface feet per minute (SFPM)

Dovetail An angled surface used on the mating parts of machine tools, a slide (dovetail slide)

Dowel A pin used to align two or more parts; may be pressed in one part with a close fit through hole in another part

Down milling Climb milling with cutter rotation in the same direction as feed

Edgefinder A tool used to locate an edge of the workpiece when setting up a job

Feed rate The amount of workpiece advance into the cutter, usually measured in inches per minute (IPM)

Feeler A thin strip of metal used to check dimensions and assist in locating cutters relative to the workpiece

Ferrous Containing iron

Forms Milling cutters that have a particular shape and are used to machine that shape (form) on a workpiece

Gang Several cutters mounted on the same arbor, all cutting at once

Gib A strip of metal in a dovetail slide located between the sliding components and capable of adjustment to minimize slide clearance

Grinding Machining by abrasive action (grinding wheels)

Hardening A heat treating process in which a metal's resistance to penetration (hardness) is increased

Helical In the form of a helix, a point on the circumference of a rotating cylinder also moving along the axis of the cylinder

Hob A special milling cutter used for special geometry such as gear teeth

Hydraulic Fluid powered; describing equipment making use of high pressure pumped fluids (hydraulic oils) to operate mechanical systems

Inserts Preshaped carbide cutting tools usually clamped in a milling cutter or other tool holder

Involute The shape of most common gear teeth in profile

Key A mechanical component used to retain cutters on arbors or other mechanical components on shafts

Knee The portion of the milling machine supporting the table and saddle and moving vertically up and down the column

Machinability The ease or difficulty of cutting a given material by the machining processes

Mechanism Transactional systems built up from mechanical components (gears, pulleys, cams, motors, linkages, etc.)

Morse One type of standard taper

N/C Numerical control; control of a machine tool and related processes by electromechanical means using numeric instructions, often employing computers

Parallel Describing lines or planes that remain the same distance apart

Periphery The area on or about the circumference of a circle; milling cutter teeth are on the periphery of the cutter

Pitch circle The centerline diameter of part features located on a circle of given size

Profiling Machining the outline of a workpiece

Plane Three or more points defined as being in or on the same surface

Pin A mechanical component used to retain parts; a cylindrical piece of metal used as a measurement aid

Rake One of the angles ground on a cutting tool

Roller A cylindrical piece of metal used as a measurement aid

Shims Small flat metal strips used to take up space under workpieces when setting up on the mill

Slab A cylindrical helical milling cutter used to surface broad areas on the workpiece

Sling A rope or cable used to rig lifts of heavy machine accessories

Spanner Any of several types of wrench usually employing pins to engage holes in the part to be turned or tightened

Spline Regularly spaced teeth cut on the circumference of a part and used to engage mating teeth on the inside or outside diameter of a mating part for the purpose of obtaining a positive drive

Stop Any of several devices set up on the milling machine and used to reference the workpiece in repeat or production machining operations

Stress Bending or distortion of the machine or work by the cut pressure

Stylus The tip of a dial indicator spindle

T-slot So called because of its shape; the inverted slot in the table of a milling machine

Taper A tool shank or part on which one end is a larger diameter than the other

Thrust A push, usually axially to a shaft, arbor, or cutter

Up mill Conventional milling, where cutter rotation opposes the direction of feed

Wiggler A tool used to locate a layout on a workpiece when setting up in the mill

Index